T0292626

Fundamentals in Organic Geochemistry

Series editors
Jan Schwarzbauer
RWTH Aachen University, Germany

Branimir Jovančićević
Faculty of Chemistry, University of Belgrade, Belgrade, Serbia

Organic Geochemistry is a modern scientific subject characterized by a high trans-disciplinarity and located at the edge of chemistry, environmental sciences, geology and biology. Therefore, there is a need for a flexible offer of appropriate academic teaching material (BSc and MSc level) addressed to the variety of students coming originally from different study disciplines. For such a flexible usage the textbook series 'Fundamentals in Organic Geochemistry' consists of different volumes with clear defined aspects and with manageable length. Students as well as lecturers will be able to choose those organic-geochemical topics that are relevant for their individual studies and programs. Hereby, it is the intention to introduce (i) clearly structured and comprehensible knowledge, (ii) process orientated learning and (iii) the complexity of natural geochemical systems. This textbook series covers different aspects of Organic Geochemistry comprising e.g. diagenetic pathways from biomolecules to molecular fossils, the chemical characterization of fossil matter, organic geochemistry in environmental sciences, and applied analytical aspects.

More information about this series at http://www.springer.com/series/13411

Jan Schwarzbauer • Branimir Jovančićević

From Biomolecules
to Chemofossils

 Springer

Jan Schwarzbauer
RWTH Aachen University
Aachen, Germany

Branimir Jovančićević
University of Belgrade
Belgrade, Serbia

ISSN 2199-8647   ISSN 2199-8655   (electronic)
Fundamentals in Organic Geochemistry
ISBN 978-3-319-27241-2   ISBN 978-3-319-25075-5   (eBook)
DOI 10.1007/978-3-319-25075-5

Library of Congress Control Number: 2016936560

This Springer imprint is published by Springer Nature
The registered company is Springer International Publishing AG Switzerland

Preface

You can approach *Organic Geochemistry* from various points of view and from different scientific disciplines due to its interdisciplinarity. However, everyone will quickly recognise a lack of knowledge in those areas that are complementary to (not covered by) his own competences. Therefore, the intention of this book series is to allow students to get familiar with Organic Geochemistry from different starting points.

In the first issue the different forms of fossil organic matter appearing in the geosphere is described from a more general point of view. This volume focusses more on the chemical aspects introducing the molecular diversity of natural products, their fate in the sedimentary systems and the consequences of the corresponding alterations for geoscientific questions. Hence, the content of this issue is organized mainly according to substance classes playing a major role in Organic Geochemistry.

Base of this textbook are lectures that are part of the geoscientific and geochemical education at RWTH Aachen University and University of Belgrade on a BSc/MSc level. Hence, this volume is addressed to students that need a detailed introduction into the chemical aspects and molecular fundamentals needed for a comprehensive understanding of Organic Geochemistry.

Aachen, Germany   Jan Schwarzbauer
Belgrade, Serbia   Branimir Jovančićević

Contents

Chapter 1
Molecular Aspects of Production and Degradation of Natural Organic Matter in the Geosphere

Natural organic matter on earth is not only abundant but also important for many essential processes. It underlies huge chemical changes, which are described by bio- and geochemical cycles. Besides modification of individual organic molecules these cycles are generally triggered by *ab initio* formation and total degradation of organic material. Therefore, the question arises, whether organic matter is principally stable or unstable under the conditions existing on earth.

1.1 Stability of Organic Matter Under Conditions as Existing on Earth

Outlook

The questions whether organic matter is stable or instable will be discussed under thermodynamic as well as kinetic aspects.

To answer this question examination of the thermodynamic stability of organic molecules might be an appropriate way. Useful parameters for estimating the stability of substances are enthalpies. Enthalpies are usually reflecting the energetic differences of educts and products of a given reaction or transformation.

Calculation of the enthalpy of formation ($\Delta H^\circ_{formation}$) can be based on a formation reaction of a certain substance from the elements in their most stable forms as educts. The enthalpy of formation is given by the differences of enthalpies of the product and the corresponding elements. Applying this approach on organic substances, some of the corresponding standard enthalpies of formation are positive, pointing to an endothermic reaction of formation and, consequently, to a higher stability of the elements as compared to the organic compounds. On the

© Springer International Publishing Switzerland 2016
J. Schwarzbauer, B. Jovančićević, *From Biomolecules to Chemofossils*,
Fundamentals in Organic Geochemistry, DOI 10.1007/978-3-319-25075-5_1

Table 1.1 Enthalpies of formation and combustion for selected organic substances

Substance	Standard enthalpy of formation $\Delta H°_{form}$ [kJ/mol]	Standard enthalpy of combustion $\Delta H°_{comb}$ [kJ/mol]
Methane CH_4 (g)	-75	-890
Acetylene C_2H_2 (g)	$+227$	-1301
Pentane, C_5H_{12} (g)	-146	-3537
Benzene C_6H_6 (l)	$+49$	-3258
Naphthalene $C_{10}H_{10}$ (s)	$+79$	-5156
Acetone CH_3COCH_3 (l)	-248	-1790
Formaldehyde HCHO (g)	-109	-571
Methanol CH_3OH (l)	-239	-726
Phenol C_6H_5OH (s)	-165	-3504
Aniline C_6H_7N (l)	$+32$	3393
Glucose $C_6H_{12}O$ (s)	-1268	-2808
Acetic acid CH_3COOH (l)	-485	-875
Glycine NH_2CH_2COOH (s)	-533	-969

contrary, some substances are characterized by negative standard enthalpies of formation, therefore, they are stable as compared to their elements. Examples are given in Table 1.1.

However, for estimating the stability of organic matter under environmental conditions the formation enthalpy might not reflect the natural situation. Since mineralization mainly to water and carbon dioxide is the degradation pathway in principle, it seems reasonable to use this reaction as a base for stability estimation. The corresponding chemical reaction is the combustion (burning with oxygen) and, therefore, enthalpies of combustion might point better to the principal stability of organic matter on the earth's surface. As illustrated in Table 1.1, for nearly all organic substances the enthalpies of combustions are negative, hence combustion of organic substances always is an exothermic reaction. Consequently, organic matter has to be characterized as thermodynamically instable under natural conditions on earth.

Then, the question arises why organic matter obviously does exist. The reason is not related to thermodynamic aspects (describing principally the state of systems) but to kinetic reasons (describing the pass from one state to another one). Every chemical reaction can be specified kinetically by the Arrhenius equation (see Fig. 1.1), in which the term 'activation energy' becomes important with respect to organic matter stability. To convert one substance into another by a chemical reaction needs such activation energy, which depends dominantly on the reaction

Fig. 1.1 The Arrhenius
equation

Arrhenius equation

$$\ln k = \ln A - E_a/RT$$

or

$$k = A * e^{-E_a/RT}$$

path and mechanism. The higher the activation energy the higher the energetic demand for initiating the reaction and the lower the overall reaction rate. That means for our question, the need to overcome the activation energy slows down the degradation of organic matter to an extent, that a metastable status can be reached.

> *General Note*
>
> Organic matter is thermodynamically unstable but kinetically stabilized. This metastable condition triggers the global carbon cycles.

1.2   Synthesis of Natural Organic Matter

> **Outlook**
>
> Photosynthesis as basic process of organic matter formation is presented with special focus on the molecular reactions at key points of this process.

Photosynthesis represent the primary process for organic matter production in nature. It is performed by autotrophic organism in a more or less similar procedure with only slight variations. The net reaction scheme is simply (see Fig. 1.2), it is the reverse reaction of mineralization, which is thermodynamically described by the enthalpy of combustion as mentioned before. Because combustion of organic matter is principally exothermic, the formation of organic substances requires

Fig. 1.2 Net reaction of
photosynthesis

$$6\ CO_2 + 6\ H_2O \longrightarrow C_6H_{12}O_6 + 6\ O_2$$

energy which is delivered dominantly by light. An exception represent a minor fraction of autotrophic organism using chemical reactions as energy sources, therefore, this organism are called chemotrophs.

Prior to discussing the molecular aspects of photosynthesis two principle biochemical systems have to be introduced briefly. The biochemical transfer of reaction energy is basically realized by the ADP/ATP system. Adenosine diphosphate (ADP) and adenosine triphosphate (ATP) differ only by one phosphate unit as illustrated in Fig. 1.3. This structural difference correlates with the energy package that can be stored or released by the reversible ADP/ATP conversion.

Secondly, a system is required for transferring hydrogen atoms as it can be deduced from the photosynthesis net reaction, since photosynthesis can also be considered as hydrogenation of carbon dioxide. Hydrogen transfer is realized in biochemical reactions by the NADPH/NADP or the NAD/NADH systems, which are illustrated in Fig. 1.4. Reversible oxidation and reduction of the nicotine amide moiety enables the adduction or release of hydrogen.

The photosynthesis is divided into two main reaction cycles, the light stage reaction cycle converting light to chemical energy and the dark stage reaction cycle

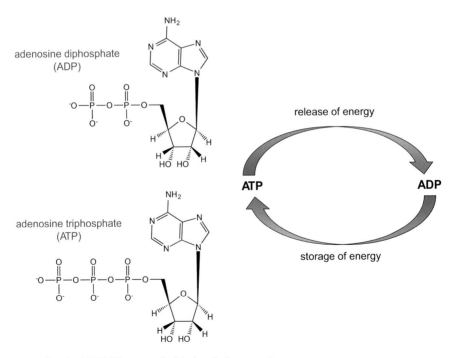

Fig. 1.3 The ATP/ADP system for biochemical energy storage

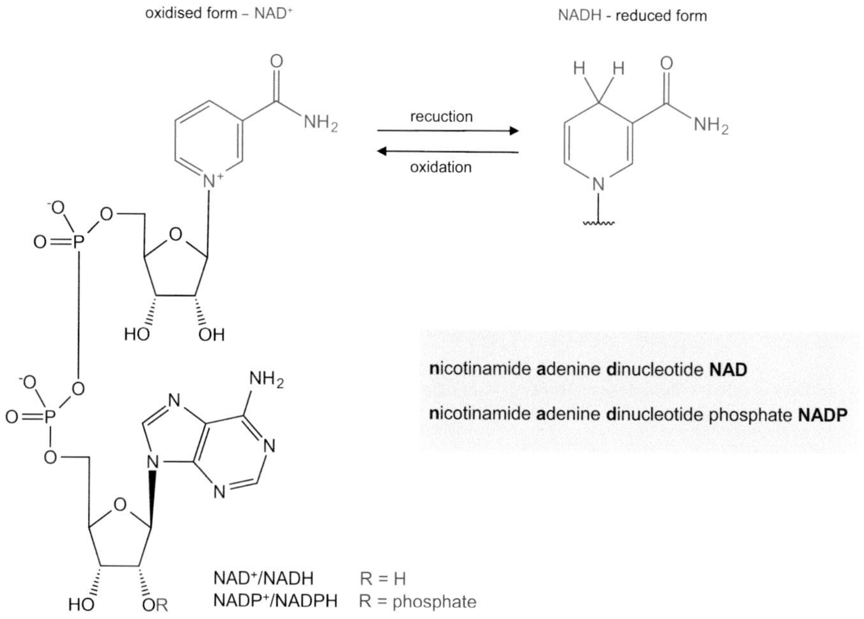

Fig. 1.4 The NAD/NADH and NADP/NADPH system for biochemical hydrogen transfer

synthesizing the organic matter. The first one uses light absorbed by chlorophyll for ATP and hydrogen formation as described in Fig. 1.5. Hydrogen is produced by lysis of water molecules under the release of oxygen. Following, the hydrogen is carried to the second cycle by the NADP/NADPH system. Hereafter, hydrogen and ATP are subjected to organic matter formation from carbon dioxide. This second cycle is commonly the so-called *calvin cycle* in which a multi-step reaction cycle incorporates CO_2 and releases glucose. The produced glucose is used for biosynthesizing all further organic biomolecules (anabolism) as well as for energy depot for catabolic reactions.

The most important step in the second cycle is the so-called CO_2 fixation, the elementary step of conversion of inorganic carbon to organic carbon. This is also the step in which some slight modifications can be seen for different species. The most common reaction is catalyzed by the enzyme *ribulose bisphosphate carbox-ylase* (RuBisCo) adding CO_2 to the ribulose 1,5-phosphate (C5 molecule) forming an unstable C6-transition state that is immediately split into two similar C3-moieties, the phospho glycerate (see Fig. 1.6). Plants using exclusively this system are called accordingly C3-plants. C3-plants represent the dominant plant species on earth.

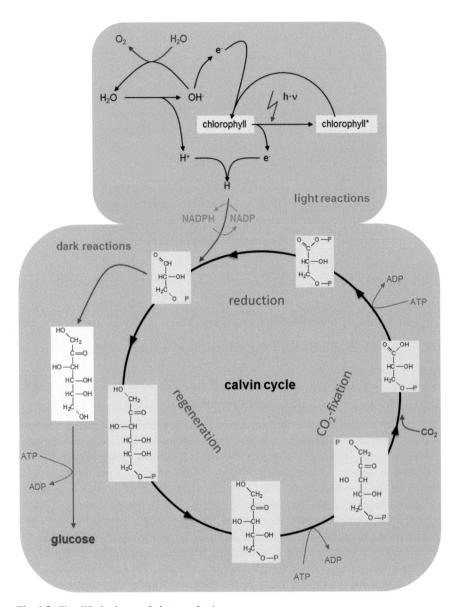

Fig. 1.5 Simplified scheme of photosynthesis

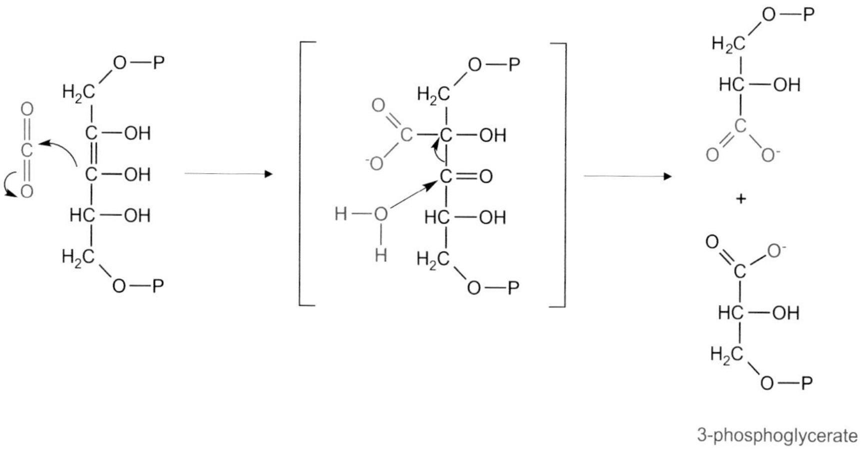

3-phosphoglycerate

Fig. 1.6 CO$_2$ fixation by RuBisCo

Alternative CO$_2$ fixation can be performed using the enzyme *phosphoenolat pyruvate carboxylase* (PEP), which in fact enables a CO$_2$ pre-fixation. Here, CO$_2$ is added to a C3-moiety, the pyruvate, forming a C4-moiety, the malate, which is temporarily used for CO$_2$ storage and for transferring it to the calvin cycle in a neighbouring cell (see Fig. 1.7). Autotrophs using this spatial separated CO$_2$ fixation mechanisms are entitled C4-plants, examples include maize or millet. Some specialized plants, in particular in arid regions, use this PEP pre-fixation too, but solely temporally separated (see Fig. 1.7). They are called CAM plants comprising e.g. crassulacea and pineapple.

Beside RuBisCo and PEP systems some further CO$_2$ fixation systems exist (e.g. reversed tricarboxylic acid cycles TCA used by some bacteria, see Table 1.2) which are of minor relevance.

General Note

Appreciable organic matter production is realized more or less exclusively by photosynthesis. With respect to the global carbon cycle, noticeable production of organic matter is located only at the earth's surface, either in the terrestrial or marine environment.

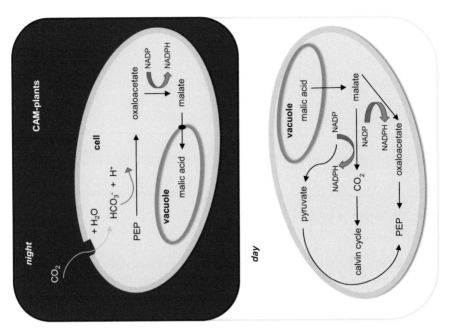

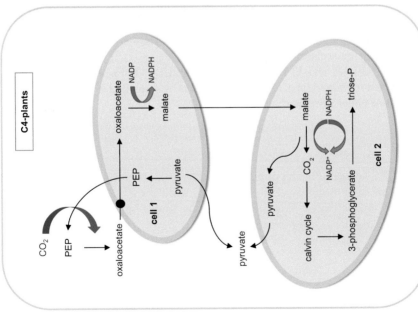

Fig. 1.7 CO_2 fixation by PEP and CAM systems

Table 1.2 Different systems of CO_2 fixation

RuBisCo	Ribulose bisphosphats carboxylate
PEP	Phosphoenolatpyruvate carboxylase
CAM	Crassulacean acid mechanism
	3-hydroxpropionate pathway
TCA	Reversed tricarboxylic acid cyclus

1.3 Degradation of Natural Organic Matter

Outlook

Degradation under aerobic and anaerobic conditions is described with respect to the processes of respiration and fermentation. The various processes are allocated to different regions in the aquatic water/sediment systems.

Processes leading to the mineralization of organic matter are not as uniform as compared to the photosynthesis as primary organic matter formation process. Since microorganisms represent by far the most active and most effective organic matter degrader, we will focus on microbial degradation processes. The structuring parameter to characterize these degradation processes is the oxygen availability, since mineralization represents an exhaustive oxidation of organic substances. Complementary, the oxidation of organic matter requires hydrogen acceptors for taking up the released hydrogen equivalents (see Table 1.3). These hydrogen acceptors become reduced as corresponding reduction reaction.

If organism have access to elemental oxygen (O_2) from air, the aerobic respiration is the most effective process for gaining energy from organic matter. Respiration in general implies complete mineralization, hence, its net reaction is the reverse reaction to photosynthesis (see Fig. 1.8, first line).

Looking on the molecular aspects of respiration generally three different steps can be distinguished. Firstly, macromolecular biomolecules such as peptides or

Table 1.3 Degradation processes under aerobic and anaerobic conditions

	Aerobic	Anaerobic
Respiration	Complete degradation → mineralization Final products: CO_2 and H_2O O_2 from air as hydrogen acceptor	Complete degradation → mineralization Final products: CO_2 and H_2O Inorganicspecies as hydrogen acceptor such as NO_3^- or SO_4^{2-}
Fermentation		Incomplete degradation Final products: e.g. ethanol, methane, lactic acid Organic substrate as hydrogen acceptor

$$C_6H_{12}O_6 \quad + \quad 6\ O_2 \longrightarrow \quad 6\ CO_2 \quad + \quad 6\ H_2O \qquad \Delta G = -2870\ kJ$$

$$C_6H_{12}O_6 \longrightarrow \quad 2\ C_2H_5OH \quad + \quad 2\ CO_2 \qquad \Delta G = -219\ kJ$$

$$C_6H_{12}O_6 \longrightarrow \quad 3\ CH_4 \quad + \quad 3\ CO_2 \qquad \Delta G = -416\ kJ$$

$$C_6H_{12}O_6 \longrightarrow \quad 2\ CH_3CH(OH)COOH \qquad \Delta G = -196\ kJ$$

Fig. 1.8 Net reaction and corresponding free standard reaction enthalpies of respiration and selected fermentation processes

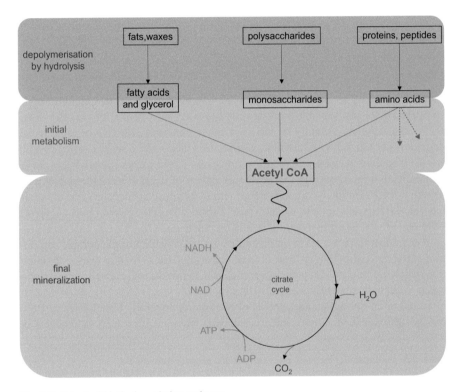

Fig. 1.9 Principal biotic degradation pathway

polysaccharides are hydrolyzed to low molecular moieties, basically pyruvate (see Figs. 1.9 and 1.12). Thereafter, transformation to activated acetic acid (acetyl coenzyme A, Acetyl CoA) by decarboxylation is performed and as the last step formation of carbon dioxide and water is gained by the so-called citrate cycle.

The *citrate cycle*, also called *Krebs cycle*, comprises several biochemical steps which results in complete oxidation of activated acetic acid to CO_2 and hydrogen equivalents stored dominantly as NADH (see Fig. 1.10). The oxidation is

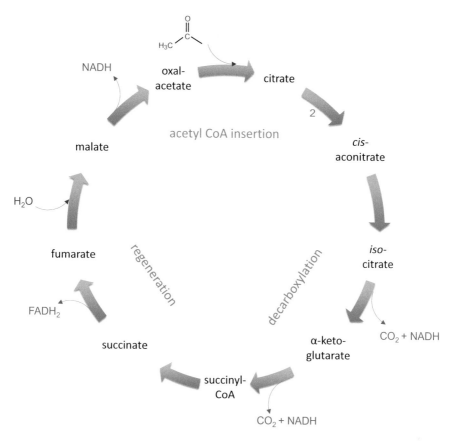

Fig. 1.10 Schematic citrate cycle

performed successively. A first reaction phase inserts acetyl CoA into the cycle by adding it to a C4 dicarboxylic acid (oxaloacetate). In a second phase decarboxylation of carboxylate moieties leads to succinate moieties which are regenerated in a third step to oxaloacetate. In the course of the total cycle, energy is gained and stored as ATP.

If free oxygen is not available, that means under anaerobic conditions, further oxygen containing reactants (e.g. sulfate SO_4^{2-}) can be used for complete oxidation and mineralization, respectively. This process is named anaerobic respiration. Anaerobic bacteria are specialized in using individual reducing reactants, most commonly sulfate and nitrate. The corresponding degradation processes are named sulfate or nitrate respiration. Less effective respiration systems in terms of energy yields are carbonate or iron respiration using either CO_2 or Fe^{3+} (see Fig. 1.11) as oxidizing agents. Anaerobic respiration is performed by either facultative anaerobes, which can switch between aerobic and anaerobic respiration, or obligatory anaerobes, that strictly live under anaerobic conditions.

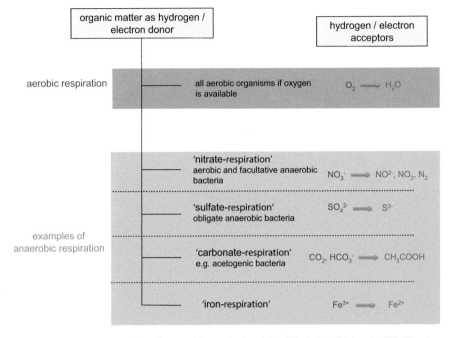

Fig. 1.11 Schematic aerobic and anaerobic respiration (simplified after Schlegel 1992; Deming and Baross 1993)

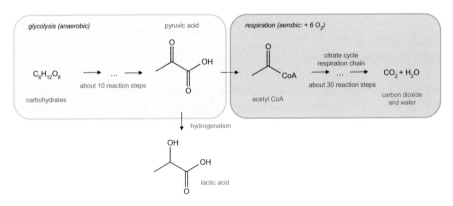

Fig. 1.12 Relation of complete and incomplete biotic degradation pathway

If only incomplete degradation is possible due to the lack of oxygen in the cells, the general degradation pathway is hindered at a stage prior to the citrate cycle. Instead of forming acetyl CoA from the precursor pyruvate, often a hydrogenation reaction forms lactic acid as final product (see Fig. 1.12). Such incomplete biotic degradation processes are called anaerobic fermentation. Alternative fermentation systems can generate further final products, e.g. alcohol fermentation produces ethanol by decarboxylation and subsequent reduction of pyruvic acid.

Incomplete degradation has a huge impact on the energy balance as compared to mineralization. The energy yield of mineralization is up to fifteen times higher as compared to fermentation as demonstrated for glucose in Fig. 1.8. Consequently, respiring species are more effective and faster in degradation of organic matter than fermenting organism. Although aerobic respiration is superior, the need of free oxygen restricts its occurrence. In aquatic sediments as well as in deeper soil layers (e.g. the saturated zone) anaerobic conditions exist and, consequently, in this ecosystems anaerobic degraders are predominant. However, there is also a hierarchy of anaerobic respiration systems controlled by their individual efficiencies and the extent of reducing conditions.

The dependence of these degradation mechanism on oxygen availability have enormous geochemical relevance since environmental compartments deliver oxygen to various extent. Following a reduction gradient (as measured by the redox potential E_h) from an aerobic water body to completely anaerobic sediment layers, the aerobic respiration in the water phase and probably on the sediment surface is replaced by nitrate respiration in the top sediment layers followed by sulfate and finally carboxylate respiration in the deeper horizons. The different zones are certainly not clearly separated but have some overlaps. Nevertheless, a clear sedimentary hierarchy of different degradation processes is obvious (see Fig. 1.13).

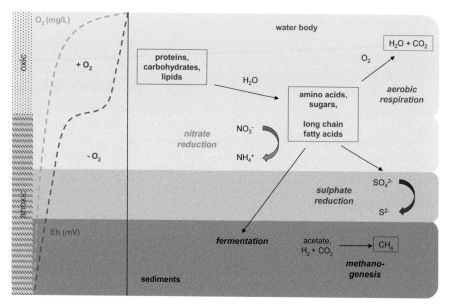

Fig. 1.13 Hierarchy of respiration and fermentation processes in water and corresponding sediments (compiled and simplified after Killops and Killops 2005; Deming and Baross 1993).

General Note

Focusing on the geochemical relevance of these different degradation processes it has to be noted, that preservation of organic matter (as required for persistence of organic matter over geological time periods) is favored under anaerobic conditions due to a less efficient and slower degradation by anaerobic microorganism.

As a last aspect, degradation of organic matter depends also on its bioavailability. Basically, two forms of organic matter exist in aquatic systems (the geochemical most relevant ecosystems) – the dissolved organic matter (DOM) and the particulate organic matter (POC). Since organism take up organic substances dominantly as dissolved species, the association to particulate matter principally protects organic compounds against biotic degradation to a certain extent. The partition between dissolved and adsorbed state is determined dominantly by the polarity of organic molecules. More lipophilic (less polar) compounds tend to adsorb on particles, whereas hydrophilic (more polar) substances remain dissolved. Since POM is more stable against microbial degradation as compared to DOM, lipophilic compounds in aquatic systems have a higher potential to persist. Further on, particle associated compounds are transported via sedimentation to the sediments, in which the more anaerobic conditions (as described) additionally protects the organic matter.

General Note

Organic matter preservation is particle associated and favored under anaerobic conditions. This is based on the low bioavailability of non-dissolved matter as well as the prevention of enhanced microbial degradation under oxygen depleted conditions. Consequently, aquatic sediments are the most appropriate ecosystems for preservation of organic matter or, with other words, sedimentary organic matter has the highest potential to persist over geological time periods.

References

Deming JW, Baross JA (1993) Chapter 5 – The early diagenesis of organic matter: bacterial activity. In: Engel MH, Macko SA (eds) Organic geochemistry – principles and applications. Plenum Press, New York/London, pp 119–144

Killops S, Killops V (2005) Introduction to organic geochemistry, 2nd edn. Blackwell Publishing, Oxford

Schlegel HG (1992) Allgemeine Mikrobiologie. Thieme Verlag, Stuttgart

Chapter 2
The Biomarker Approach

Outlook

Organic molecules altered by diagenetic processes partially allow to obtain information about paleoenvironmental conditions as well as sedimentary processes after deposition. Biomolecules keeping such specificity or indicator properties even after their conversion to chemofossils are entitled 'biomarker'. They are able to point to the original organic matter sources or to characterize either the depositional conditions or the thermal maturity of the fossil matter.

From a historical point of view a major force driving the scientific field of Organic Geochemistry was the linkage of biological molecules with corresponding molecular constituents in fossil material like oil and coals. Contemporarily, this relation was also the initial point of this scientific matter. In the beginning 1930s the knowledge about the origin of petroleum was very limited although its practical value and usefulness in particular as energy resource was well recognized. A report by Alfred E. Treibs published 1936 in the journal *Angewandte Chemie* (see Fig. 2.1) pointed for the first time to a close relationship between plant material as obvious biological material and certain oil constituents. The specific molecular structure of porphyrins, in particular the assembly of nitrogen containing five membered rings (see Sect. 5.3), detected in oil shales represents the molecular residues of porphyrin containing biomolecules, especially chlorophyll which characterizes perfectly plant material. Till now many so-called chemofossils or biomarkers have been identified, discussed and linked to biogenic precursors. In the meanwhile, biomarker analysis is an established tool not only for solving scientific problems but also for application in oil industry, in particular for exploration business.

© Springer International Publishing Switzerland 2016
J. Schwarzbauer, B. Jovančićević, *From Biomolecules to Chemofossils*,
Fundamentals in Organic Geochemistry, DOI 10.1007/978-3-319-25075-5_2

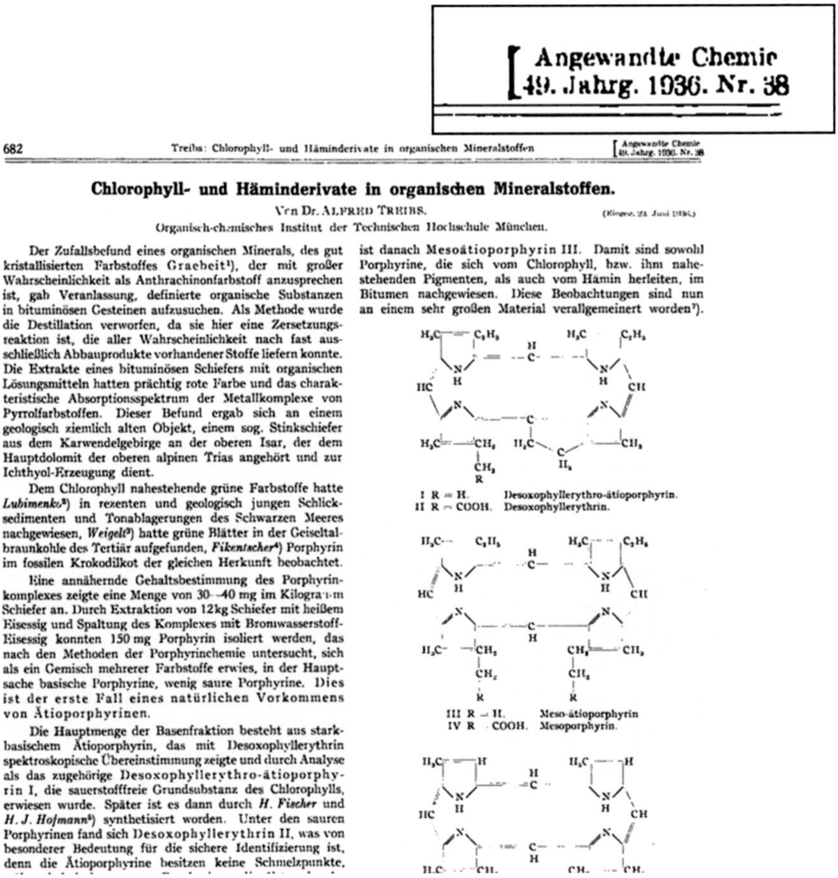

Angewandte Chemie
49. Jahrg. 1936. Nr. 38

682 Treibs: Chlorophyll- und Häminderivate in organischen Mineralstoffen [Angewandte Chemie 49. Jahrg. 1936. Nr. 38

Chlorophyll- und Häminderivate in organischen Mineralstoffen.

Von Dr. ALFRED TREIBS. (Eingeg. 24. Juni 1936.)

Organisch-chemisches Institut der Technischen Hochschule München.

Der Zufallsbefund eines organischen Minerals, des gut kristallisierten Farbstoffes Graebeit[1]), der mit großer Wahrscheinlichkeit als Anthrachinonfarbstoff anzusprechen ist, gab Veranlassung, definierte organische Substanzen in bituminösen Gesteinen aufzusuchen. Als Methode wurde die Destillation verworfen, da sie hier eine Zersetzungsreaktion ist, die aller Wahrscheinlichkeit nach fast ausschließlich Abbauprodukte vorhandener Stoffe liefern konnte. Die Extrakte eines bituminösen Schiefers mit organischen Lösungsmitteln hatten prächtig rote Farbe und das charakteristische Absorptionsspektrum der Metallkomplexe von Pyrrolfarbstoffen. Dieser Befund ergab sich an einem geologisch ziemlich alten Objekt, einem sog. Stinkschiefer aus dem Karwendelgebirge an der oberen Isar, der dem Hauptdolomit der oberen alpinen Trias angehört und zur Ichthyol-Erzeugung dient.

Dem Chlorophyll nahestehende grüne Farbstoffe hatte *Lubimenko*[2]) in rezenten und geologisch jungen Schlicksedimenten und Tonablagerungen des Schwarzen Meeres nachgewiesen, *Weigelt*[3]) hatte grüne Blätter in der Geiseltalbraunkohle des Tertiär aufgefunden, *Fikentscher*[4]) Porphyrin im fossilen Krokodilkot der gleichen Herkunft beobachtet.

Eine annähernde Gehaltsbestimmung des Porphyrinkomplexes zeigte eine Menge von 30–40 mg im Kilogramm Schiefer an. Durch Extraktion von 12 kg Schiefer mit heißem Eisessig und Spaltung des Komplexes mit Bromwasserstoff-Eisessig konnten 150 mg Porphyrin isoliert werden, das nach den Methoden der Porphyrinchemie untersucht, sich als ein Gemisch mehrerer Farbstoffe erwies, in der Hauptsache basische Porphyrine, wenig saure Porphyrine. Dies ist der erste Fall eines natürlichen Vorkommens von Ätioporphyrinen.

Die Hauptmenge der Basenfraktion besteht aus starkbasischem Ätioporphyrin, das mit Desoxophyllerythrin spektroskopische Übereinstimmung zeigte und durch Analyse als das zugehörige Desoxophyllerythro-ätioporphyrin I, die sauerstofffreie Grundsubstanz des Chlorophylls, erwiesen wurde. Später ist es dann durch *H. Fischer* und *H. J. Hofmann*[5]) synthetisiert worden. Unter den sauren Porphyrinen fand sich Desoxophyllerythrin II, was von besonderer Bedeutung für die sichere Identifizierung ist, denn die Ätioporphyrine besitzen keine Schmelzpunkte,

ist danach Mesoätioporphyrin III. Damit sind sowohl Porphyrine, die sich vom Chlorophyll, bzw. ihm nahestehenden Pigmenten, als auch vom Hämin herleiten, im Bitumen nachgewiesen. Diese Beobachtungen sind nun an einem sehr großen Material verallgemeinert worden[7]).

Fig. 2.1 The original publication by Alfred E. Treibs

Organic substances from dead organism or segregated by living biota are normally subjected to rapid mineralization, but a very small fraction remains preserved in sedimentary systems. Principal conditions for preservation are already introduced in Chap. 1. As we have seen preservation of organic matter is preferred (i) in aquatic systems, (ii) particle associated and (iii) under anaerobic conditions, in summary in subaquatic sediments. Towards its way to these regions of preservation or geoaccumulation organic matter is subjected to degradation to various extent in the different zones of the aquatic systems. This is illustrated in Fig. 2.2. In the productive aquatic zone, the photic zone, organic matter is *ab initio* synthesized and mineralized but also segregated by phototrophic organism. Consumption and segregation of organic matter by heterotrophic organism represents a second food cycle. Segregated organic matter contributes to the dissolved organic matter (DOM), but adsorption of less polar compounds transfers a part of the dissolved matter to a particulate bound state, namely the particulate organic matter POC.

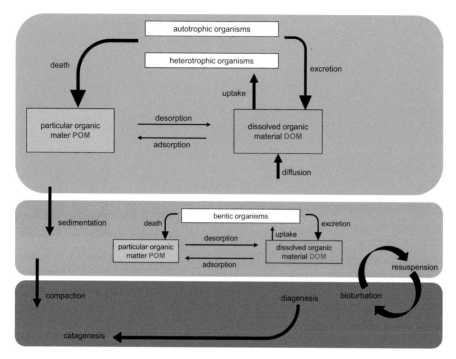

Fig. 2.2 Principal cycles of organic matter conversion in aquatic systems (compiled, modified and simplified after Killops and Killops 2005; Deming and Barros 1993; Wakeham and Lee 1993)

Noteworthy, this process is reversible comprising contemporary adsorption and desorption. Beside organic matter segregated by living organism, detritus derived from dead organism contributes to both DOC and POC.

However, the transfer process by adsorption involves two effects. Firstly, adsorbed substances are less bioaccessible (as already introduced in Chap. 1) and, therefore, not directly available to re-enter the food cycle. Secondly, under appropriate conditions particles underlie sedimentation implying a transport of the associated organic matter towards the deeper water layer and the sediments. However, in the benthic zone the reversibility of adsorption and desorption causes a second cycle of intake and excretion of the corresponding organic matter, in this zone by benthic organism. With ongoing sedimentation the particulate associated organic matter elude from higher benthic organism but microbial attacking remains. This phase represents the inception of the so-called diagenesis comprising all (mostly biotic) alterations in sediments under low temperature and low pressure conditions (see Fig. 2.3).

From a molecular point of view the diagenetic changes of organic matter can be outlined as illustrated in Fig. 2.4. The primary residues as segregated from living organism or derived from necrotic biomass consist dominantly of biogenic macromolecular matter derived from polysaccharides, peptides etc. However, also low molecular weight molecules such as lipids survive. During diagenesis the organic

Fig. 2.3 Differentiating
characteristics of diagenesis
and catagenesis

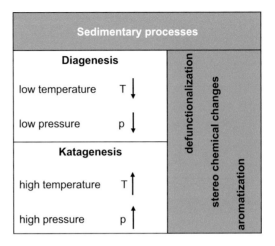

matter is transformed to humic substances and later on to kerogen. Within this
matter, low molecular weight substances can be embedded and partly protected, in
particular against further microbial degradation. However, these low molecular
weight molecules did not survive unaltered. Chemical alterations can be attributed
roughly to some principal reaction types comprising defunctionalization, aromatization and stereochemical changes like epimerization (see Fig. 2.3). In principal,
such diagenetic modifications are initiated by microorganism and lead to the
formation of more stable substances (in terms of thermodynamical stability
and/or less biodegradability). Consequently, the potential to persist increases.

General Note

Diagenetic impact on organic substances lead to the formation of
defunctionalized and stabilized molecules, finally to pure aliphatic and/or
aromatic hydrocarbons.

Further on, the transformation of individual molecules can be categorized and
roughly attributed to the different stages of diagenesis and catagenesis (see
Fig. 2.3). The latter one follows the diagenesis and is characterized by dominantly
abiotic reactions under higher temperature and pressures, thus is dominantly triggered by achieving higher thermodynamical stability

Considering all this aspects for preferential persistence over geological time
scales, selected substance classes appear to be preferred for reoccurrence in fossil
matter. In particular lipids tend to accumulate in sediments (especially in the
particulate associated organic matter) and they are less biodegradable or more
recalcitrant as compared to many other higher functionalized and more polar biomolecules. Therefore, many biomarkers belong to the group of lipids.

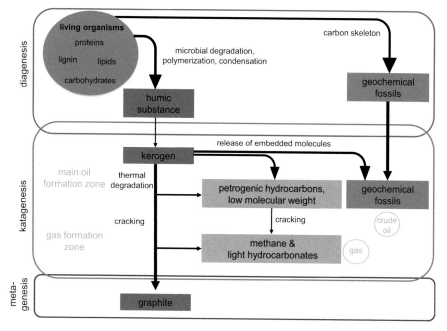

Fig. 2.4 Chemical fate of organic matter in the geosphere (compiled, modified and simplified after Tissot and Welte 1984; de Leeuw and Largeau 1993; Tegelaar et al. 1989)

But what are the characteristics, the meaning and the requirements of 'biomarkers'? Basic idea of the biomarker approach is to use the relationship between biogenic precursor substances and the fossil molecules comprising the knowledge of the diagnostic transformation pathway to gain information about diagenetic processes, the paleoenvironmental conditions as well as the composition of the original biological material (see Fig. 2.5).

---

1. Paleoenvironmental information

a. Original composition of biogenic material e.g. terrestrial plant contribution marine organisms, bacteria, etc.

  → molecular structure

b. Depositional conditions e.g. facies, paleo-temperature, CO_2-partial pressure, redox-conditions, etc.

  → variation in diagenetic pathways, diagenetic products and isotropic shifts

---

2. Sedimentary processes

a. Thermal maturity

  → thermodynamic stability vs. biogenic signature

Fig. 2.5 Principal objectives of biomarker applications

There are preconditions for using chemofossils as biomarker that are depending on the field of its application. Firstly, characterization of *original composition of biogenic material* (part 1a in Fig. 2.4) needs a linking of biomarker and biogenic precursor as unambiguous as possible. Such a high source specifity implies a high similarity of the molecular structure of precursor and diagenetic product. This correlates well with application of morphological fossils used to reconstruct paleoenvironments. The basic idea of chemofossils (or chemical biomarker) is oriented to classical paleontological research as illustrated roughly in Fig. 2.6.

Appropriate structural properties includes structural as well as stereochemical isomerism. An explicit example is related to a biomolecule named β-amyrine, which is a well-known constituent in angiosperms (see Fig. 2.7). Angiosperms represent an important group of higher land plants. In fossil matter a corresponding biomarker molecule can be identified called oleanane. A comparison of the molecular structures indicates a loss of some features like a double bond or the hydroxyl group (processes of defunctionalization, see Fig. 2.3). However, the complete carbon skeleton remains unaffected as well as the stereochemical orientation of four chiral atoms with methyl substitution (indicated by bonds noted as lines or

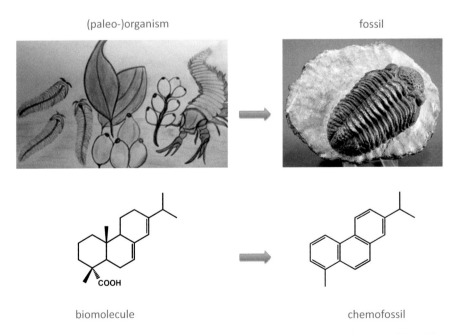

Fig. 2.6 Correlation of fossils and chemcofossil with respect to their origins (copyright of the fossil figure: KRAUS Georeproduktion 2014, reprint with kind permission)

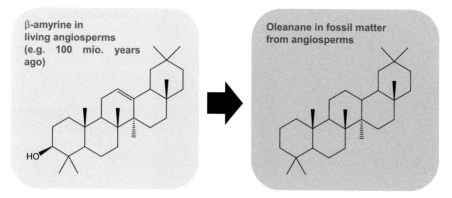

Fig. 2.7 The structural relationship of the biomolecule β-amyrine as indicator for angiosperms and its chemofossil oleanane

wedges in the molecular formula in Fig. 2.7). Hence, a very close relationship between biomolecule and chemofossil based on the chemical structure is obvious. Consequently, oleanane can act as angiosperm indicator in fossil matter.

General Note

Specific or even unique structural moieties surviving under diagenetic and catagenetic stress are the key points for chemofossils to act as valuable biomarker.

Noteworthy, such assignments need caution since biomarker might derive from different biogenic precursors. This is exemplified in Fig. 2.8 presenting biomarker molecules derived from steroids. Steroids and their chemofossils demonstrate the limitations and the varying quality of biomarker in terms of validity. A well working example is related to dinosterol and its diagenetic product dinosteranes. This steroid is exclusively produced by dinoflagellates, a monocellular phytoplankton occurring widespread in the aquatic environment, and exhibit a unique methylation at 4-position as compared to other common steroids. This structural feature remains in the dinosterane making this 4-methyl steranes to appropriate biomarkers indicating the contribution of dinoflagellates to the fossil organic matter. Looking on other steranes, an unambiguous linkage of the biomarker molecules to individual steroids is not always feasible. The loss of functional groups and unsaturated moieties leads to structural identical chemofossils for the biogenic steroids (i) cholesterol and desmosterol or (ii) brassicasterol and campesterol or (iii) fucosterol, stigmasterol and β-sitosterol, respectively (see Fig. 2.8). However, the corresponding fossil steranes point clearly to steroids as precursor but unfortunately

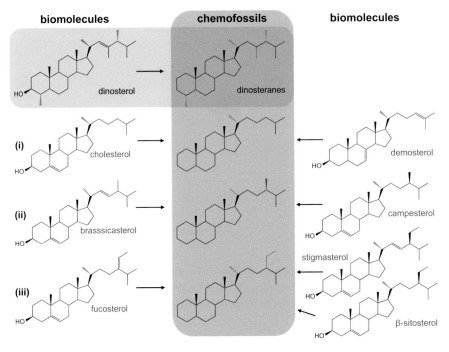

Fig. 2.8 Different source specifities of steranes as compared to their biological precursors

do not characterize individual sterols, hence their information value is lower as compared to dinosteranes.

General Note

The higher the structural similarity of biomarker precursor and corresponding chemofossil the higher the biomarker quality.

Accordingly, the biogenic source specifity of some substances classes are summarized illustrating their usefulness and precision to act as biomarker. Some substances or compound group are pointing to biota groups (e.g. to angiosperms, bacteria) but others much more precise to individual organism (dinoflagelates). Some compound classes acting as biomarker and their corresponding biogenic sources are summarized in Fig. 2.9.

Secondly, to obtain *information about paleo conditions and sedimentary processes* (part 1b and 2 in Fig. 2.5) the diagenetic pathway need to be known and to be understood accurately. One important field of application of biomarkers is the determination of the maturity level of fossil matter. Molecular changes during diagenesis but also during catagenesis (or from the biomolecule towards the

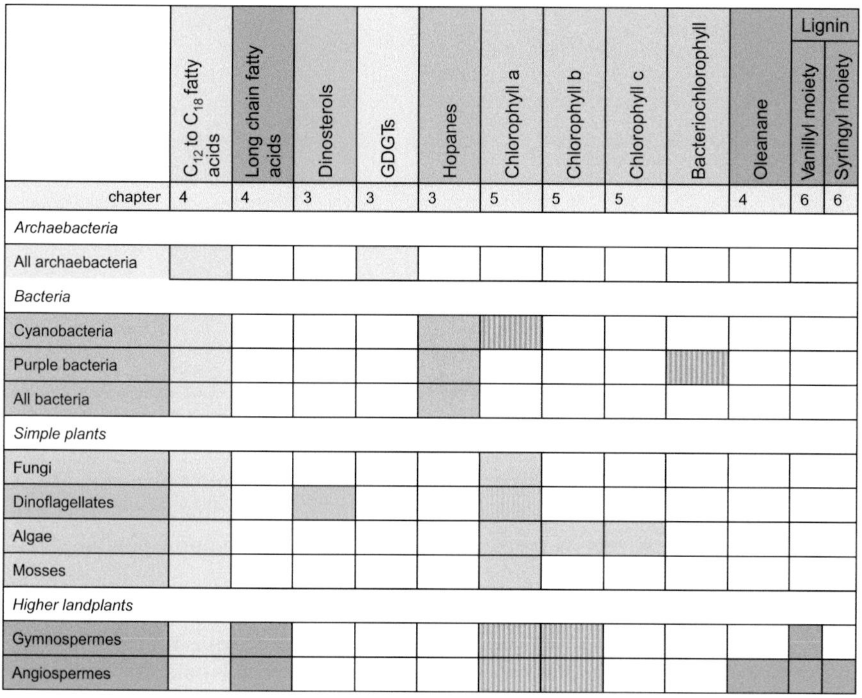

	C$_{12}$ to C$_{18}$ fatty acids	Long chain fatty acids	Dinosterols	GDGTs	Hopanes	Chlorophyll a	Chlorophyll b	Chlorophyll c	Bacteriochlorophyll	Oleanane	Lignin Vanillyl moiety	Lignin Syringyl moiety
chapter	4	4	3	3	3	5	5	5		4	6	6
Archaebacteria												
All archaebacteria												
Bacteria												
Cyanobacteria												
Purple bacteria												
All bacteria												
Simple plants												
Fungi												
Dinoflagellates												
Algae												
Mosses												
Higher landplants												
Gymnospermes												
Angiospermes												

Fig. 2.9 Correlation of indicative biomarker groups and corresponding source organism

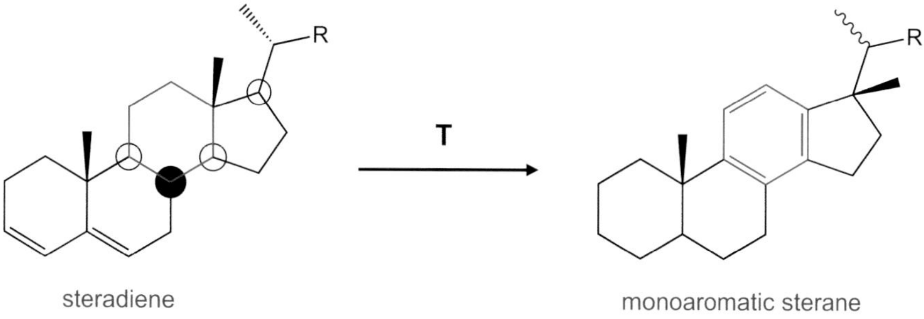

steradiene   T   monoaromatic sterane

Fig. 2.10 Aromatization as reaction step during sterine diagenesis

chemofossil) are triggered generally by an increasing thermodynamical or chemical stability. The individual steps within one diagenetic/catagenic pathway are more or less consecutive and the stage of transformation is dominantly related to the maximum temperature and, consequently, the maturity. Further on, maturity correlates certainly also with the maximum burial depth.

An example of this correlation is given in Fig. 2.10. During the geochemical transformation of steroids an aromatization of the C-ring is observable. The formation of this thermodynamical more stable aromatic system is provoked by

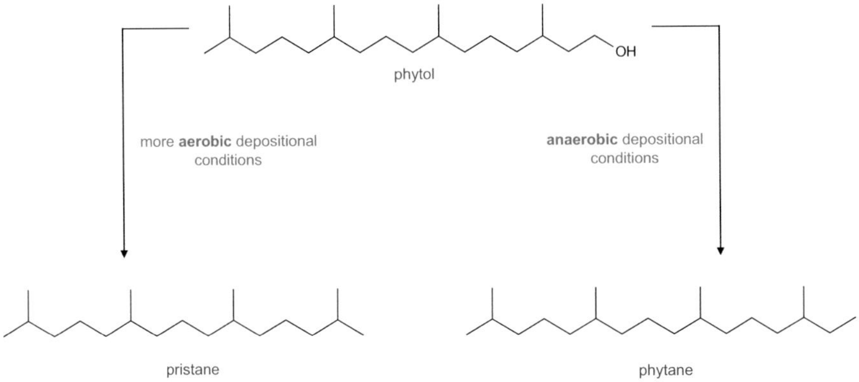

Fig. 2.11 Simplified diagenetic pathway of phytol under different sedimentation conditions

temperature and, therefore, an increasing relative proportion of the aromatic as compared to the nonaromatic derivatives follows ongoing thermal stress. This ratio can be used quantitatively to predict thermal maturity of the corresponding fossil matter.

If processes at the very early stage of diagenesis are influenced by the environmental conditions, the resulting changes in biomarker composition can be used for tracing the paleoenvironmental status. This is another type of biomarker application revealing *information about paleo conditions and sedimentary processes* (part 1b and 2 in Fig. 2.5) and is illustrated in Fig. 2.11 with a very well-known organic geochemical example. Briefly, phytol derived dominantly from plant chlorophyll and its diagenetic pathway is very well investigated. The production of the two main diagenetic products pristane and phytane vary according to the availability of oxygen during the early stage of sedimentation. Consequently, differences in the quantitative relation of these products reflect diverse paleoenvironmental conditions. Roughly summarized, the higher the pristane/phytane ratio the more aerobic conditions existed during sedimentation.

General Note

Chemical variations as result of different environmental conditions during deposition or sedimentation can reveal paleoenvironmental information. Chemical alteration during diagenesis and catagenesis can point quantitatively to the thermal maturity of the organic matter.

As shown by all these examples using the biomarker approach needs two skills, a profound knowledge about the biological precursors (their molecular structures, origin and relationship to organism) as well as a detailed insight into the diagenetic and catagenetic reactions these biogenic molecules were subjected to. Hence, in the following chapters biological molecules with relevance for organic geochemistry will be introduced. In particular, their molecular and physico-chemical properties, their physiological relevance and their diagenetic pathways will be discussed in detail. Some examples of corresponding biomarker application will complement the different sections.

References

De Leeuw JW, Largeau C (1993) A review of macromolecular organic compounds that comprise living organisms and their role in kerogen, coal, and petroleum formation. In: Engel MH, Macko SA (eds) Organic geochemistry – principles and applications. Plenum Press, New York/London, pp 23–72

Deming JW, Baross JA (1993) Chapter 5 – The early diagenesis of organic matter: bacterial activity. In: Engel MH, Macko SA (eds) Organic geochemistry – principles and applications. Plenum Press, New York/London, pp 119–144. ISBN 0-306-44378-3

Killops S, Killops V (2005) Introduction to organic geochemistry, 2nd edn. Blackwell Publishing, Oxford. ISBN 0-632-06504-4

Tegelaar EW, Derenne S, Largeau C, de Leeuw JW (1989) A reappraisal of kerogen formation. Geochim Cosmochim Acta 3:3101–3107

Tissot BP, Welte DH (1984) Petroleum formation and occurrence, a new approach to oil and gas exploration. Springer-Verlag, Berlin, Heidelberg, 1978. Second Revised and Enlarged Edition

Wakeham SG, Lee C (1993) Chapter 6 – Production, transport and alteration of particulate organic matter in the marine water column. In: Engel MH, Macko SA (eds) Organic geochemistry – principles and applications. Plenum Press, New York/London, pp 145–170. ISBN 0-306-44378-3

Further Reading

Brocks JJ, Pearson A (2005) Building the biomarker tree of life. Rev Mineral Geochem 59:233–258

Hunt JM, Philp RP, Kvenvolden KA (2002) Early developments in petroleum geochemistry. Org Geochem 33:1025–1052

Simoneit BRT (2004) Biomarkers (molecular fossils) as geochemical indicators of life. Adv Space Res 33:1255–1261

Chapter 3
Isoprenoids

The compound class of isoprenoids exhibits three groups with geochemical relevance: the terpenoids, the hopanes and the steroids. As a fourth group also GDGTs exhibit highly relevant structural units of isoprenoid origin. All these substance groups are linked via their biosyntheses, which are using primarily the same basic structural educt, the 2-methyl-1,3-butadiene or isoprene (see Fig. 3.1). Principally, isoprenoids are formed by the repetitional addition of these isoprene units.

Biotic formation of isoprene uses activated acetic acid units, the so-called acetyl CoA, which are coupled thrice to form activated mevalonic acid (see Fig. 3.2). After reduction and decarboxylation the isoprene is obtained as phosphate conjugate, a form of activated isoprene. Since mevalonic acid is an important intermediate in biosynthesis, it is called mevalonic acid pathway MVA and terpenoids are also known as mevalogenines. It has to be noted, that an alternative synthesis route is used by plants and bacteria to a minor extent, the MEP/DOXP or non-mevalonate pathway. Nevertheless both pathways end up with the same products, two activated isomers of isoprene.

Noteworthy, due to its regular methyl substituents, isoprenoic substances are easy to identify by their typical regular methyl group substitution along the main carbon-skeleton. This high degree of branching induces an elevated probability for asymmetrically substituted carbon atoms and, consequently, for chirality. Hence, most of the terpenoids and steroids exhibit not only one chiral carbon atom but also often a higher number of these chiral centers (partially more than 10). As a consequence multiple stereoisomers, enantiomers and diastereomers exist for

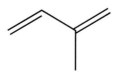

Fig. 3.1 Isoprene (2-methyl-1,3-butadiene)

© Springer International Publishing Switzerland 2016
J. Schwarzbauer, B. Jovančićević, *From Biomolecules to Chemofossils,*
Fundamentals in Organic Geochemistry, DOI 10.1007/978-3-319-25075-5_3

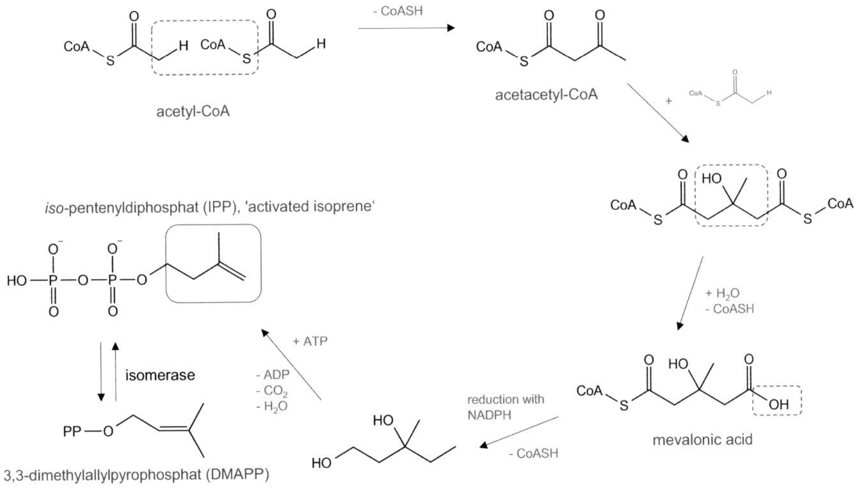

Fig. 3.2 Scheme of isoprene biosynthesis

most of the isoprenoid compounds. These stereochemical properties play an important role in Organic Geochemistry, since geoscientific information can be gained from both the alteration as well as the preservation of the stereochemistry of biomolecules.

3.1   Occurrence, Structure and Physico-chemical Properties of Mono- to Tetraterpenes

Outlook

Principal structural properties and basic biological functions of terpenes are described.

Terpenes and terpenoids are highly abundant substance classes in nature, in particular in plants, with a wide spectrum of biological functions. These functions are related to the physico-chemical properties of the compounds, in particular to volatility, polarity and steric demand. Furthermore, chemical reactivity also contributes to the type of biological function.

The group of terpenoids can be sub classified by different structural aspects. Terpenoids appear naturally not only as pure hydrocarbons, the so-called terpenes, but also as functionalized compounds such as alcohols, ethers, aldehydes and ketones. Further on, terpenoids can be subdivided into the groups of mono-, di-,

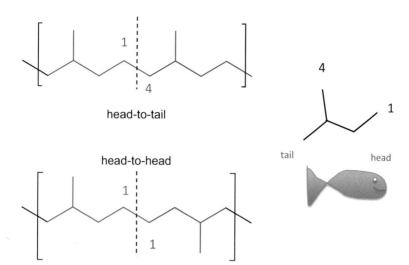

head-to-tail

head-to-head

Fig. 3.3 Principal linkages of isoprene units in terpenes

tri- and tetraterpenes as well as sesqui- and sesterterpenes according to the number of isoprene units they are built of. Lastly, acyclic, alicyclic and aromatic compounds can be differentiated. All these structural differences determine basically the physico-chemical properties of the individual terpenoids, such as boiling point, vapor pressure, chemical reactivity, steric properties etc.

Generally, the discrimination of mono- to tetraterpenes is the most common classification. The chemical linkage of the individual isoprene units generating the different terpene groups follows strict rules. These rules are based on the differentiation of both ends of the isoprene molecule labelled with 'head' and 'tail' according to Fig. 3.3. In nature the connection between two isoprenoic units are restricted to *head to tail* (1–4 linkage) or *head to head* linkages (1–1 linkage, see Fig. 3.3).

Excursus: *Where is the head and where is the tail?*

This nomenclature has one problem. Head and tail annotation can be changed (see below), and the tail, as described here, becomes the head. However, the general rules certainly remain valid in principal, but with reversed terminology.

Head and tail annotation as used by L. Ruzicka, the discoverer of the *isoprene rule* (this annotation is also used in this book)

(continued)

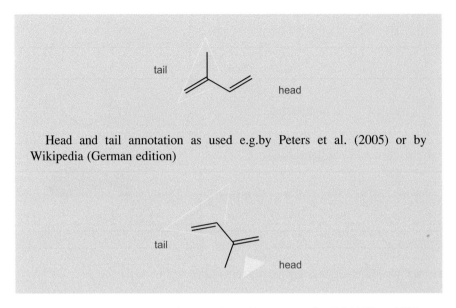

Head and tail annotation as used e.g.by Peters et al. (2005) or by Wikipedia (German edition)

Monoterpenes are build up of two activated isoprene units (DMAPP and IPP, see Figs. 3.2 and 3.4) resulting in carbon skeletons with ten atoms linked *head to tail*. Primary product of the monoterpene synthesis is activated geranol (GPP). Sesquiterpenes and diterpens are biosynthesized in the same mode of linkage (*head to tail*) by a following addition of one or two further isoprene moieties to the GPP unit. This results in activated farnesol (FPP) or geranylgeranol (GGPP) with 15 and 20 carbon atoms, respectively. A change of linkage can be observed for tri- and tetraterpenes, which are built by two sesqui- or diterpenoic units (FPP or GGPP) with *head to head* orientation, respectively (see Fig. 3.4). The principal constitution comprising number of isoprene moieties, corresponding number of carbon atoms and types of linkages are summarized in Table 3.1 for mono- to tetraterpenes and illustrated in Fig. 3.4.

General Note

Isoprenoidal compounds are easily to identify among the huge spectra of biomolecules. Their unique structural element is the frequent methyl substitution with its very regular order of substitution positions.

3.1.1   Monoterpenes

Monoterpenes represent an important group of plant constituents. Due to their low molecular weight and the corresponding high volatility as well as their aromatic flavors they are used as fragrances (especially in blossoms and fruits) and

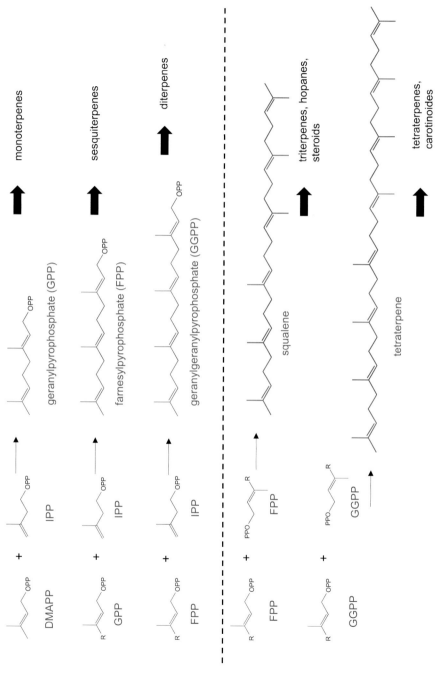

Fig. 3.4 Biosynthetic construction path of mono- to tetraterpenes using the activated moieties IPP, DMAPP, FPP and GGPP

Table 3.1 Principal composition and structure of major terpene subclasses

Substance class	Number of isoprene units	Number of carbon atoms	Type of linkage
Monoterpenoids	2	10	*Head-to-tail* of isoprene units
Sesquiterpenoidss	3	15	*Head-to-tail* of isoprene units
Diterpenoids	4	20	*Head-to-tail* of terpene units
Sesterterpenoids	5	25	
Triterpenoids	6	30	*Head-to-head* of sesquiterpene units
Tetraterpenoids	8	40	*Head-to-head* of diterpene units

pheromones. Further on, they represent main ingredients of essential oils like peppermint, lemon or eucalyptus oil as well as turpentine. Compound names of terpenoids are often not systematic according to UIPAC but reflect their biological occurrence. Examples include geraniol, limonene or citronellol. The molecular properties comprise acyclic, cyclic, aromatic carbon backbones as well as several functionalities, e.g. hydroxyl or carboxylic groups. As mentioned above numerous monoterpenes exhibit chiral centers. Some examples of common monoterpenes are illustrated in Fig. 3.5.

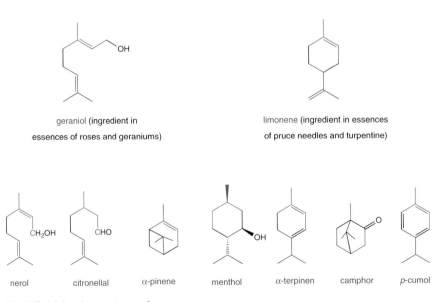

geraniol (ingredient in
essences of roses and geraniums)

limonene (ingredient in essences
of pruce needles and turpentine)

nerol   citronellal   α-pinene   menthol   α-terpinen   camphor   p-cumol

Fig. 3.5 Molecular structures of some common monoterpenes

3.1.2 Sesquiterpenes

Following the systematic formation route of terpenoids the addition of one more isoprene unit leads to the group of sesquiterpenes. Their biological functions as well as the principal structural properties are similar as compared to monoterpenes. However, a slight trend to a higher degree of cyclization is obvious. Examples include guajazulene and cadinene (Fig. 3.6).

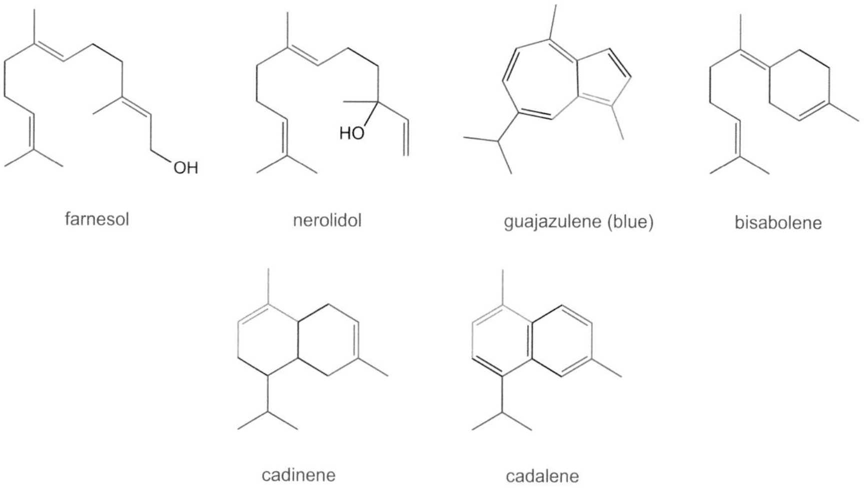

farnesol   nerolidol   guajazulene (blue)   bisabolene

cadinene   cadalene

Fig. 3.6 Molecular structure of common sesquiterpenes

Excursus: *Terpenoids as archeological indicators*

Terpene analyses applied to ancient pottery jars have been used to get information about the composition of Egypt wines (McGovern et al. 2009). Monoterpenes revealed information about potential herbal additives such as rosemary, mint, coriander, thyme and some more (see table below). The identification of retene, dehydroabietic acidand a few more cyclic diterpenes indicated the use of pine resin or corresponding tar as coating of the amphora. This resins contributed certainly also the overall taste of old Egyptian wine.

Table: Terpenes identified in amphores from Abydos and Djerbel Adda according to McGovern et al. (2009). Additionally, potential biogenic sources of terpenes are listed

Terpene components	Possible Sources
p-cymol	pine, rosemary
fenchon	rosemary, fennel, sage
α-terpineole	pine, mint, wine
carvon	mint, yarrow, sage, artemisia
vanillin	rosemary, thyme
farnesole	pine
biformen	pine

3.1.3   Diterpenes

Due to their higher molecular mass and the corresponding lower volatility the biological function of diterpenes change to the usage as regulatory agents, vitamins and ingredients of plant resins (examples are given in Fig. 3.7). An example for a physiological highly relevant diterpene is retinol, better known as vitamin A, which is an essential and vital nutrient for humans.

From an organic-geochemical point of view two different diterpenes are highly valuable in particular for biomarker studies (see Fig. 3.8). Firstly, the tricyclic diterpene abietic acid is an important constituent of conifer resins in which this compound (among other structurally related tricyclic acids) is responsible for the gumming process following the excretion of resins to protect surface lesion of barks. Abietic acid is one of the best investigated compounds in Organic Geochemistry, since the diagenetic products of this molecule appear in many fossil samples and act as indicator for conifer derived organic matter.

A second important diterpene with respect to organic-geochemistry is phytol, an acyclic alcohol. Phytol is dominantly a constituent of chlorophyll, but to a minor extent also of other biomolecules. Nevertheless, the appearance of phytol and

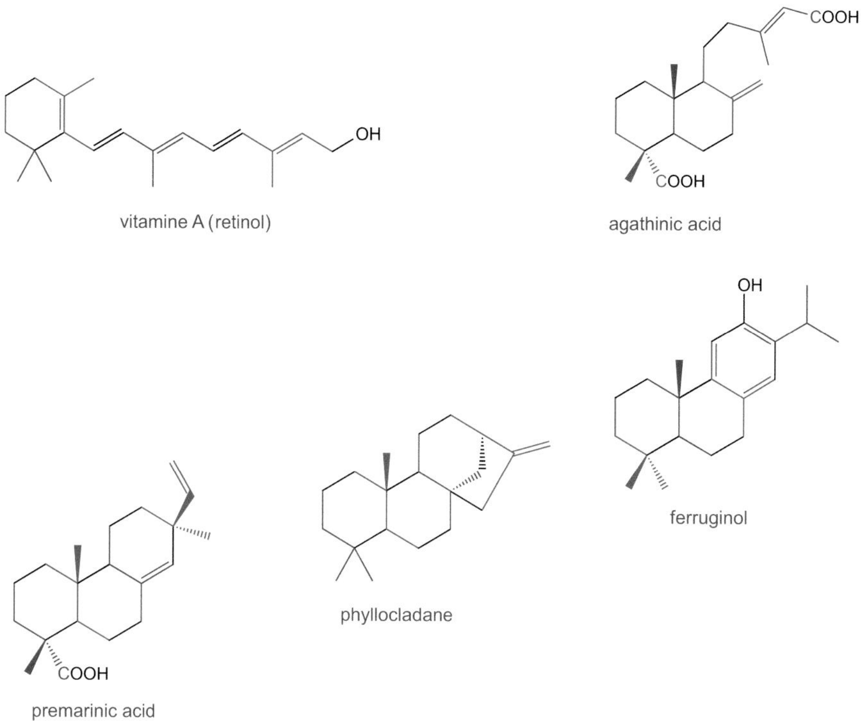

Fig. 3.7   Molecular structures of some common diterpenes

especially of its diagenetic products normally indicates the contribution of photosynthesizing organisms to the organic matter.

Due to the high significance of abietic acid and phytol in Organic Geochemistry their diagenetic pathways will be discussed in separate subchapters. Since phytol is closely related to the plant pigment chlorophyll, its geochemical fate and biomarker quality will be presented in the chapter pigments.

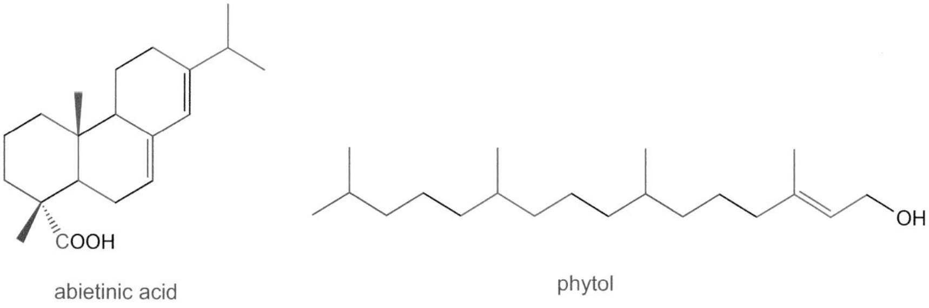

Fig. 3.8   Diterpenes of high geochemical importance

3.1.4   Triterpenes

Triterpenes appear in nature dominantly as cyclic compounds. In plants pentacyclic triterpenes act as bittern, protection and resistance agent and represent those constituents that have a high importance in Organic Geochemistry. Some examples are given in Fig. 3.9. Due to the high degree of branched moieties in the molecules numerous chiral or stereogenic centers exist.

From a structural point of view, the cyclization of the acyclic precursor squalene results in three different types of basic structures for pentacyclic triterpenes. In nature, variations are observable at the E-ring covering ring size, length of aliphatic substituents and substitution positions. In principal, oleanane-, ursane- and lupine-type skeletons occur (see Fig. 3.10).

One of the best organic-geochemically investigated pentacyclic triterpenes is amyrin which exists as α- or β-isomer (see Fig. 3.11). Amyrins are indicative substances for angiosperms and gymnosperms. Due to its geochemical significance the diagenetic pathway of β-amyrin is discussed in detail in Sect. 3.2.2.

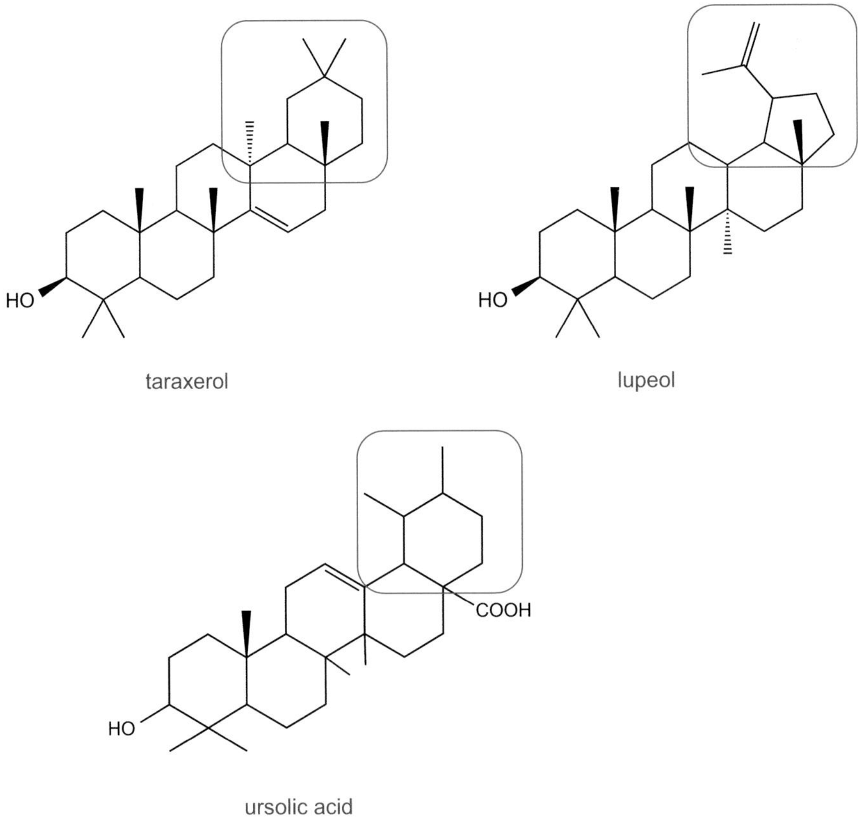

taraxerol   lupeol

ursolic acid

Fig. 3.9 Molecular structure of some common triterpenes

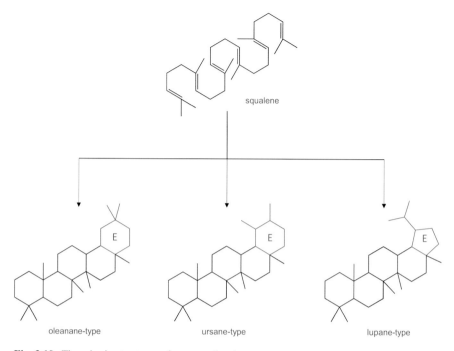

Fig. 3.10 Three basic structures of pentacyclic triterpenes

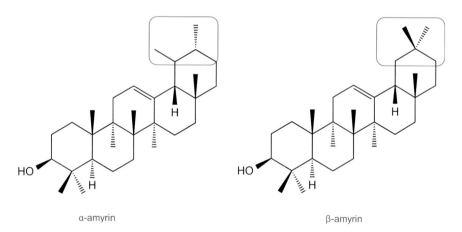

Fig. 3.11 Triterpenes of high geochemical importance

3.1.5 *Tetraterpenes*

The relevance of tetraterpenes in Organic Geochemistry is restricted as compared to the smaller terpenoids. However, one group of structural and biogenetic related substances, the carotenoids, is not only of high biological relevance but has also

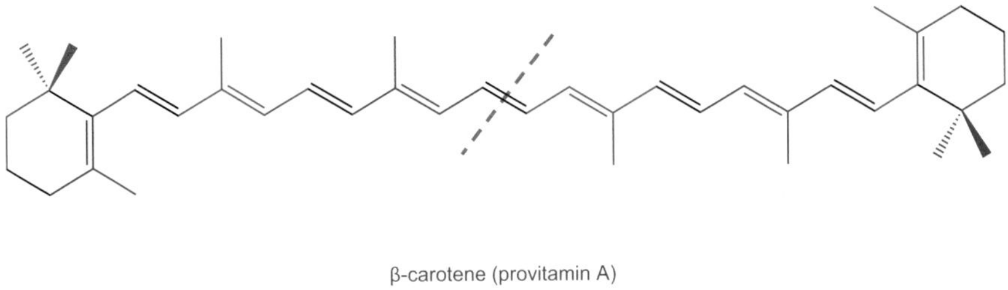

β-carotene (provitamin A)

Fig. 3.12 Carotene as very well known tetraterpene

been subjected intensively to biomarker analyses. Beside the best known derivative β-carotene, also known as provitamin A (see Fig. 3.12), numerous carotenoid isomers serve as pigments in terrestrial as well as marine plants. Details of the geochemical fate and relevance of carotenoids are discussed in the Chap. 5 – *pigments*.

General Note

A tendency to elevated cyclisation for higher terpenes (from mono- to tetraterpenes) is obvious.

3.2 Selected Diagenetic Transformations

Outlook

On two distinctive examples the changes of the molecular structures due to diagenetic and catagenetic reactions are discussed in detail. Both examples represent differences in diagenetic pathway and its complexity, but point to some basic transformations namely defunctionalization, aromatization and stereochemical changes.

3.2.1 *Abietic Acid*

As already mentioned, abietic acid is one of the best investigated biomarkers. Beside its source specifity also the diagenesis and the corresponding diagenetic derivatives are of general interest. The diagenetic pathway follows a simple reaction scheme (see Fig. 3.13). Firstly, the two double bonds situated in the ring system are used for the formation of an aromatic C-ring by further dehydrogenation. The so-called dehydroabietic acid is generated. The order of the following reaction steps can vary. Either the removal of the functional group by decarboxylation, forming an

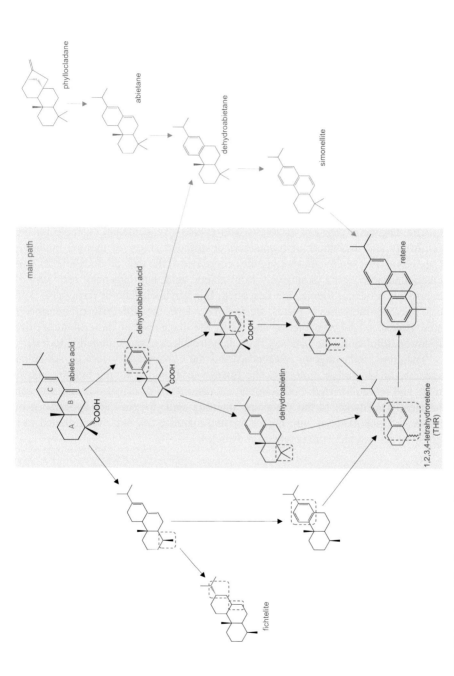

Fig. 3.13 Major steps of the diagenetic pathway of abietic acid (compiled and simplified after Otto and Simoneit 2002; Killops and Killops 2005 Hautevelle et al. 2006)

unsaturated hydrocarbon (dehydroabietine) is followed by the aromatization of ring B. Or the defunctionalization occurs after a first dehydrogenation, introducing a new double bond. Then, the complete aromatization is obtained by a further dehydrogenation step. Both different pathways lead to the same diagenesis product, the tetrahydroretene (THR). As the last step the third ring is dehydrogenated and contemporarily demethylated to form a phenanthrene system with the typical 7-methyl and 3-isopropyl substitution. This so-called retene is the thermodynamically stable end product of the diagenetic pathway of abietinic acid, in which the ring system, one methyl substituent and the isopropyl group are preserved. On the contrary, some structural information and properties have been lost, comprising (i) the stereochemical configuration at three bridging ring carbon atoms, (ii) the carboxyl and one methyl group attached to a bridging carbon atom as well as (iii) two double bonds and information on their position. In summary, the diagenetic pathway of dehydroabietic acid is characterized dominantly by defunctionalization and aromatization forming a thermodynamically stable aromatic hydrocarbon.

The diagenetic pathway of abietic acid is also a fundamental example for the sensitivity and usefulness of biomarkers at different stages of thermal maturity. Care has to be taken by interpreting the occurrence of the final product, retene. This compound derived not exclusively from abietic acid, but diagenetic pathways of alternative biogenic sources are described. E.g. phyllocladane or ferruginol, both also tricylcic diterpenes in plants, are forming retene, but with different intermediates. Instead of dehydroabietine and THR, the diagenesis of these biomolecules followed dominantly the pathway from dehydroabietane and simonellite to retene.

Therefore, with ongoing diagenesis and the contemporary loss of structural specifity of biomarker molecules the linkage to a unique source biomolecule is more and more restricted. For example, the occurrence of dehydroabietic acid can be clearly attributed to the precursor abietinic acid, whereas the appearance of retene in fossil material with higher thermal maturity can only be attributed to a group of several biogenic precursor molecules.

Case Example: Organic geochemical application of abietic acid based biomarker

Retene as final diagenetic product of abietic acid has been used among other high plant indicators by van Aarsen et al. (2000) to follow paleovegetation changes. They reflected the increase of higher land plant vegetation during Jurassic time in three Australian sediment cores. As a fingerprint they did not analyse solely retene but accompanied by other indicators such as cadalene. Summing up the relative contributions of these parameters defined a more general 'higher plant parameter' HPP. The relative abundance of HPP was correlated successfully with paleoclimatic conditions and global sea level.

Excursus: *Abietic acid, colophonium and retene*

Resins of conifers are technically treated by water steam distillation in order to obtain turpentine oil. The residue of this technical process is called colophonium, which consists of different tricyclic diterpene acids. Abietic acid is only a minor constituent of colophonium, but treatment with soft acids (e.g. acetic acid) transfers all other diterpene acids into abietinic acid. This process is applied to produce abietic acid on a technical scale. In particular the salts of abietinic acids salts are used as detergents and glue with high relevance in the paper industry.

Abietic acid can be synthetically aromatized by heating with sulphur resulting in the same final product as compared to the diagenetic and catagenetic transformation. This reaction of retene forming has been used to elucidate the molecular structure of abietinic acid.

abietinic acid   retene

3.2.2   Amyrin

The pentacyclic triterpene amyrin can also act as an excellent example illustrating all the different aspects of diagenetic pathways. In contrast to abietinic acid, the transformation from the biomolecule α- or β-amyrin to stable biomarkers is more complex. It consists of various pathways and final products. However, some general aspects of the diagenetic fate of biomolecules can be deduced exemplarily.

The first diagenesis pathway is characterized by defunctionalization resulting in the hydration of the double bond and the reduction of the hydroxy-group (see Fig. 3.14). Finally, a pure aliphatic hydrocarbon, the so-called oleanane, remains with a characteristic polycyclic core structure in which many stereochemical specification (at rings A, B and C) of the biomolecule β-amyrin persisted. Stereochemical changes are evident at ring D due to double bond rearrangement and epimerization at the final reversible reaction step converting 18β-oleanane to

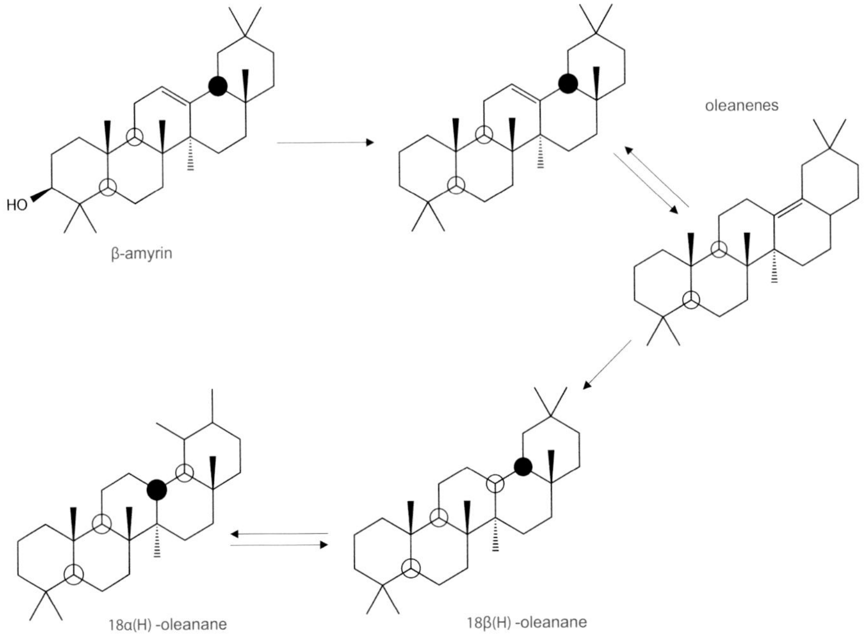

Fig. 3.14 Schematic diagenetic pathway of β-amyrin forming oleanane (simplified after Killops and Killops 2005)

18α-oleanane. In summary, due to the persistence (i) of the characteristic polycyclic structure, (ii) of the specific methyl substitution pattern and, in particular, (iii) of many stereochemical properties, the biomarker oleanane exhibit a high biomarker potential. Principally, the occurrence of oleanane in fossil matter is indicative for higher land plants.

Case Example: Occurrence of oleanane and fossil record

How oleanane can act as indicator for angiosperms has been depicted impressively by Moldowan (1994) A clearly visible correlation of micro fossils (pollen of angiosperms) with the detection frequency and amount of the corresponding chemofossil demonstrates the high biomarker quality of oleanane as angiosperm indicator. Noteworthy, the occurrence of oleanane beginning in the Jurassic followed by a starting appearance of angiosperm pollen in the Cretaceous points to the idea, that 'chemical evolution' might be somewhat antecedent as compared to the biological evolution.

(continued)

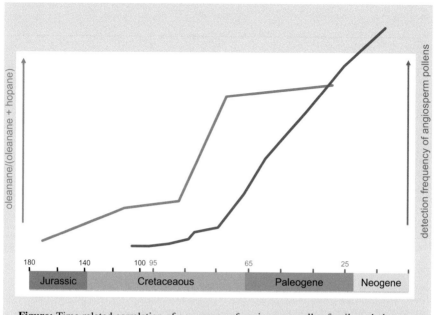

Figure: Time related correlation of occurrence of angiosperm pollen fossils and oleanane detection (normalized to hopane amount) in rock samples (simplified after Moldowan 1994)

An alternative diagenetic route is linked to aromatization processes that are promoted by the preformed cyclic structures. As illustrated in Fig. 3.15 aromatization starts for α-amyrin at ring A followed by further aromatization with the order B > C > D > E. During aromatization various methyl substituents (methyl substituents at bridging carbon atoms or dimethyl substituents at germinal positions) split off and the final product is a substituted five ring aromatic hydrocarbon, 1,2,9-trimethylpicene. Under certain circumstances the A-ring gets lost during the initial diagenesis and the following aromatization leads finally to a four ring aromatic hydrocarbon, the 3,6,7-trimethylchrysene. For both products a limited source specificity in terms of biomarker application has to be stated. This is related to the loss of more indicative structural properties as compared to the diagenetic formation of oleanane.

A third diagenetic pathway of β-amyrin has a more destructive initial reaction step (see Fig. 3.16). The C-ring can be cleaved leading to a destabilization of the polycyclic system. Accordingly, after aromatization of the sub ring systems, the resulting derivatives tend to break down into two separate naphthalene or tetraline moieties with various methyl substitution patterns. These final products exhibit only very low indicative properties and remain less specific as compared to the other diagenetic products described above.

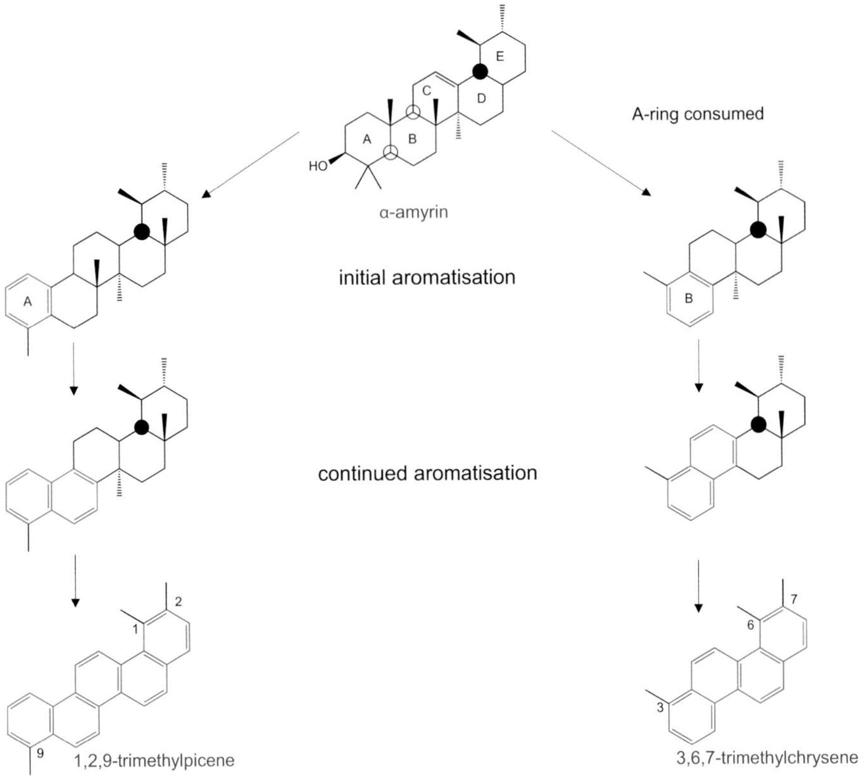

Fig. 3.15 Schematic diagenetic pathway of α-amyrin forming polycyclic aromatic hydrocarbons (simplified after Killops and Killops 2005 and references cited therein)

Generally, the described diverse diagenetic fates of amyrins clearly demonstrate the complexity of diagenetic pathways and, contemporarily, the competition of different transformation routes. A principal summary for the diagenesis of amyrins is given in Fig. 3.17, which is also valid for many more, particularly cyclic, biomolecules. The final products can act with different quality as biomarker. Often, the structural features are preserved best in aliphatic hydrocarbon bio-markers. Therefore, these biomarkers are more suitable for biological source characterization as compared to the alternative aromatic hydrocarbon end products. However, due to the linkage of aromatization with thermal stress, these compounds are used frequently as thermal maturity indicators. Lastly, the higher the loss of structural information during a diagenetic pathway the lower the usefulness of the final products to act as biomarker. This is exemplified for the decomposition products during the third diagenetic transformation route of amyrin.

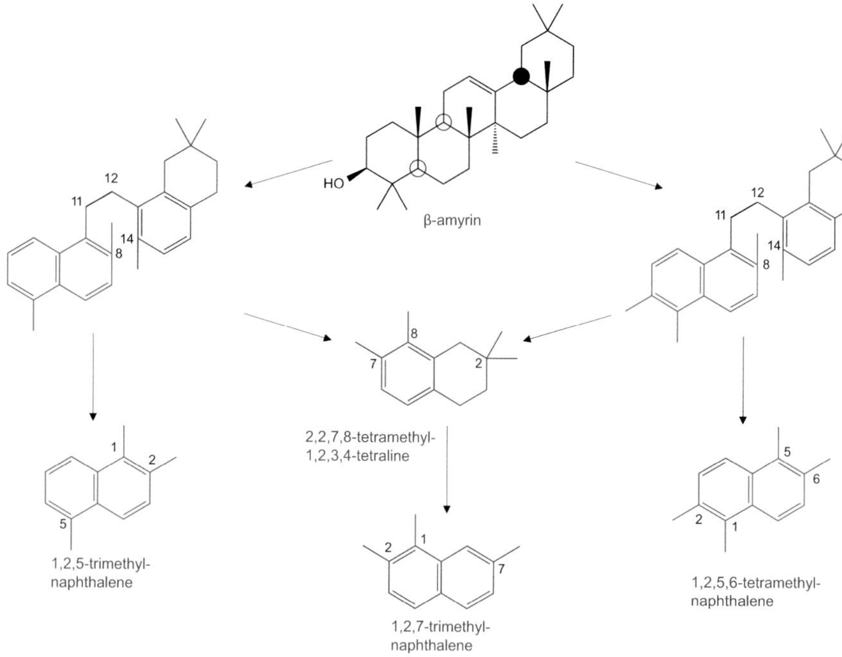

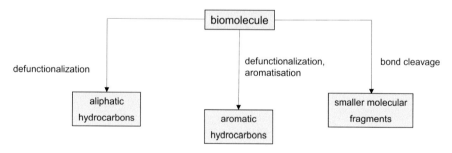

Fig. 3.16 Schematic diagenetic pathway of β-amyrin forming naphthalene derivatives as break down products (simplified after Killops and Killops 2005 and references cited therein)

Fig. 3.17 General scheme of basic diagenetic pathways forming stable biomarkers

General Note

Most recalcitrant diagenetic and catagenetic endpoints of terpenoids are aliphatic and aromatic hydrocarbons retaining structural properties of the biogenic precursors to a different extent.

3.3   Occurrence and Structural Properties of Steroids

Outlook

As a specific sub class of isoprenoids the group of steroids are discussed with respect to their molecular structure, their biological occurrence and function as well as their principal biosynthesis.

In Organic Geochemistry a very important class of biomolecules and their molecular fossil are the steroids. All steroids exhibit the same molecular core as illustrated in Fig. 3.18. Four aliphatic ring systems (three six- and one five-membered rings) are condensed in alternating linear and angular directions. The rings are all linked normally in *trans* configuration forming a more or less planar molecule.

Excursus: *Steroids as fecal indicators*

The connection of steroid rings is not restricted exclusively to *trans* configuration. One exception is used in environmental studies as follows: The hydrogenation of the common cholesterol leads to cholestanol, in which the connection of the A and B ring is newly defined during this reaction. The transformation in nearly all animals forms the *trans* linked product, but human intestinal bacteria generate the *cis* linked product named coprostanol. Due to the more or less unique human origin of coprostanol this compounds are used as indicators for human fecal contamination. Coprostanol and related derivatives (e.g. coprostanone) are also called fecal steroids.

Structural variations building up the diversity of steroids are located at only a few substitution positions marked as R1 and R2 in Fig. 3.19. Generally, three subclasses are characterized: (i) bile acids, (ii) hormones and (iii) sterines. They differ in the number of carbon atoms, where the smallest ones are the hormones with 18–21 carbon atoms. Well known representatives are testosterone or estrogen, important hormones in human organism. Further examples are given in Fig. 3.20.

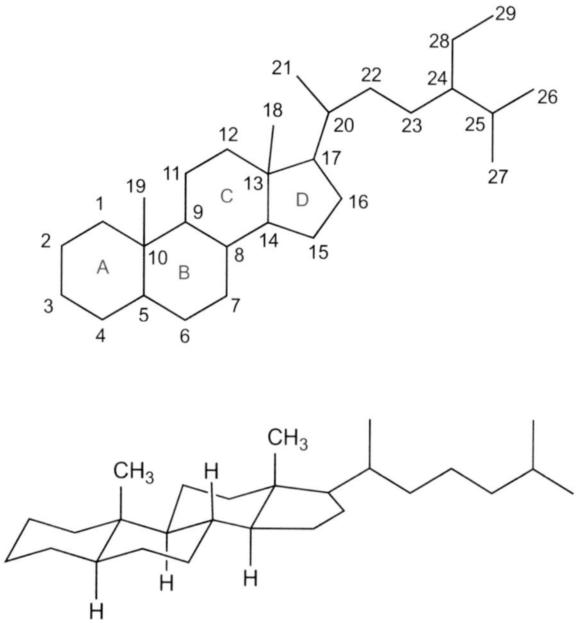

Fig. 3.18 Basic structure of steroids

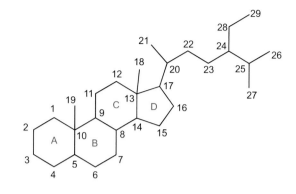

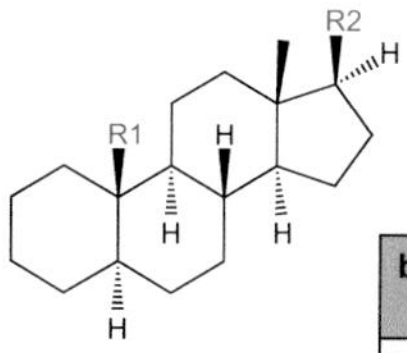

Fig. 3.19 Systematic structural diversity of steroids

basic structure	R1	R2		
5α-estrane (C$_{18}$)	H	H		
5α-androstane (C$_{19}$)	CH$_3$	H		hormones: C$_{18}$ – C$_{21}$
5α-pregnane (C$_{21}$)	CH$_3$			
5α-cholane (C$_{24}$)	CH$_3$			bile acids: C$_{24}$
5α-cholestane (C$_{27}$)	CH$_3$			
5α-ergostane (C$_{28}$)	CH$_3$			sterines: C$_{27}$ – C$_{29}$
5α-stigmastane (C$_{29}$)	CH$_3$			

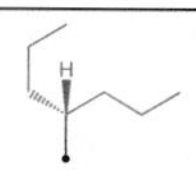

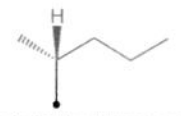

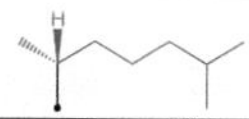

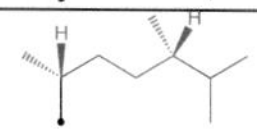

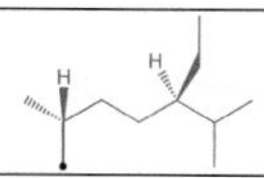

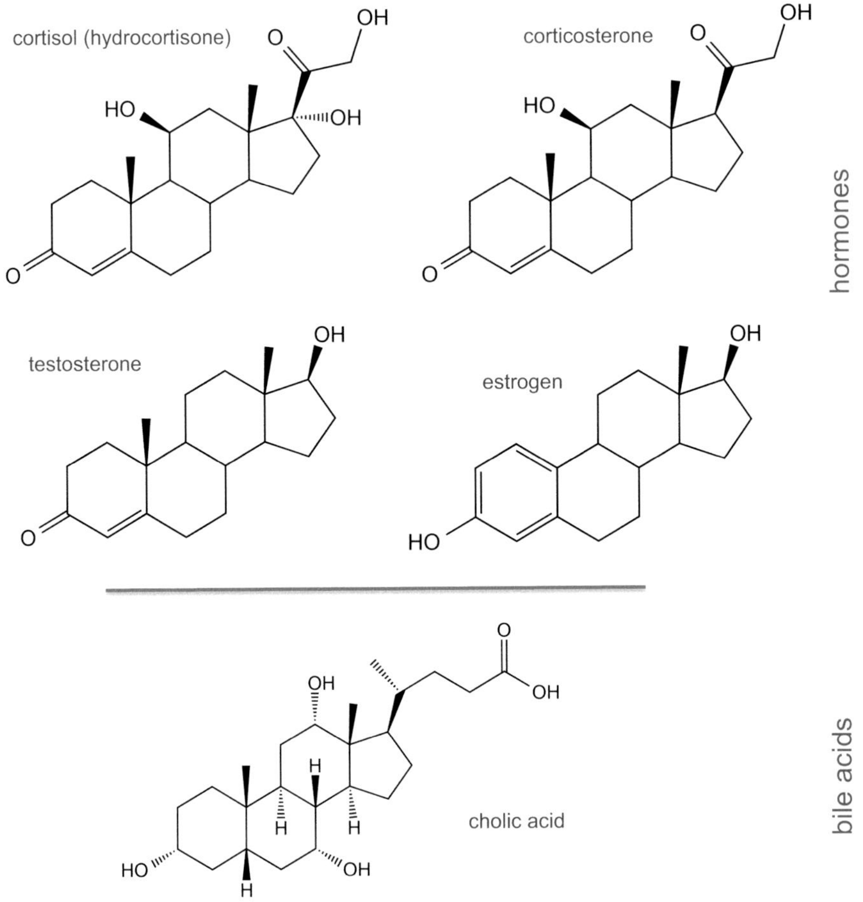

Fig. 3.20 Examples of bile acids and hormones

Bile acids consist of 24 carbon atoms and are produced in bile of higher organisms (e.g. cholic acid, lithocholic acid).

> **_General Note_**
>
> Solely the length of the aliphatic side chain in steroids determines the biological function (hormone, bile secretion, cell membrane constituent) in higher organism.

However, geochemical importance is limited to the sterines and their diagenetic products. These compounds are important components of cell membranes of eucaryotes. *Inter alia* they act as modulator for micro-scale fluidic properties of

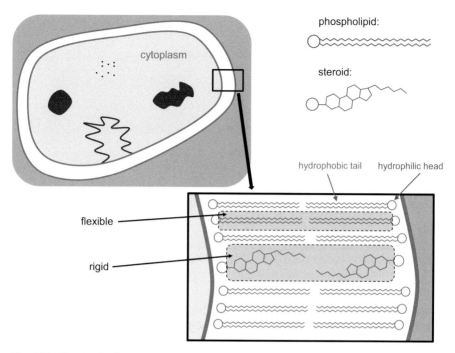

Fig. 3.21 Sketch of cell membrane structure of procaryotes

membranes (see Fig. 3.21). Sterines exhibit a 27–29 carbon atom skeleton. The biosynthesis of these compounds is based on the tetraterpene squalene, which is activated in a first step by epoxidation forming a three membered oxygen-containing ring with high chemical reactivity (see Fig. 3.22).

Cleavage of the epoxid initiates a cascade of ring formation resulting in the specific four-ring system of sterines (three six membered and one five membered ring all linked in *trans*-configuration, see Fig. 3.18). At this stage the biosynthesis of C27/28 and C29 sterines diverges. Intermediate in the first sub pathway is lanosterol, which is finally converted to steroids like cholesterol (C27) and ergosterol (C28) (see Fig. 3.21). The latter biosynthesis forms cycloartenol as intermediate which is finally converted to the so-called phytosterines, typically with 29 carbon atoms. Well known phytosteroids are stigmasterol and sitosterol.

Beside the slightly different biosynthesis pathways, the three groups of sterines appear differently in the biosphere (see Fig. 3.23). Cholesterol for example is widely distributed and occurs in nearly all organisms (plants, animals etc.). The phytosterols stigmasterol and sitosterol are constituents dominantly of higher land plants, ergosterol is a typical ingredient of fungi and brassicasterol appears in several unicellular algae. A very specific example, dinosterol (with an unusual methyl substitution at the C4 position at ring A) can be attributed primarily to dinoflagellates, a form of marine diatoms.

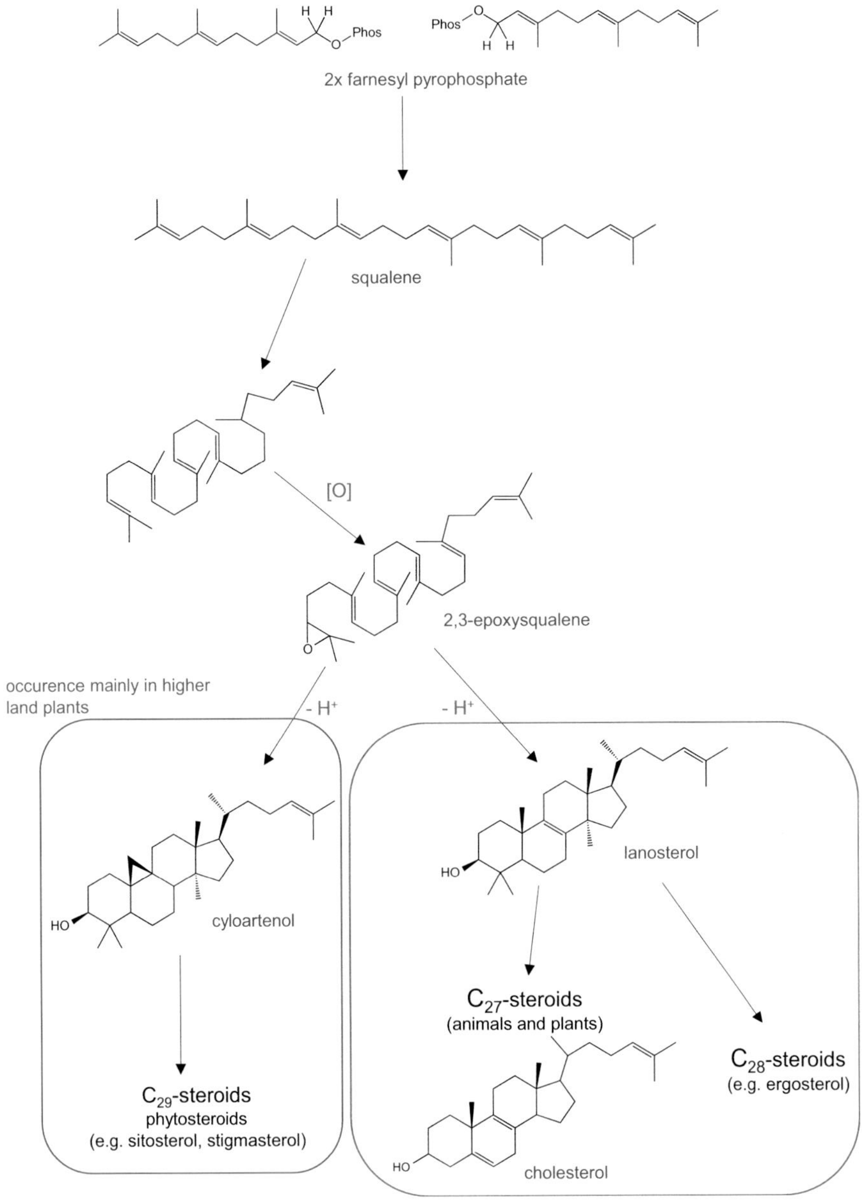

Fig. 3.22 Biosynthesis of steroids (according to Volkmann 2005; Summons et al. 2006; Killops and Killops 2005)

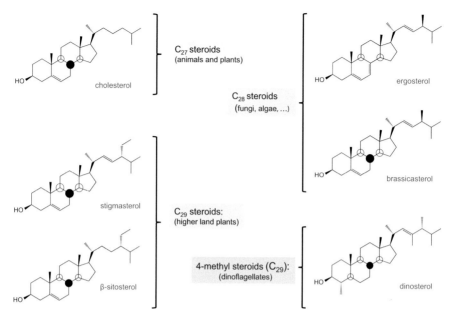

Fig. 3.23 Examples for compounds of some main groups of sterins

3.4   Diagenetic Pathways of Steroids and Their Application as Biomarker

Outlook

Also steroids underlie a complex diagenetic alteration, which is presented in detail here. In addition, some approaches using steroid related biomarkers for assessing either paleoenvironmental conditions or thermal maturity are discussed exemplarily.

As pointed out above, some biogenic sterines exhibit a high potential to act as indicators for groups of organism or even species. The structural differences of individual sterines are related dominantly to the occurrence and position of double bonds as well as to stereochemical properties (connection of rings, orientation of substituents at chiral carbon atoms in the noncyclic part of the molecules). For a geochemical usage of the indicative properties it becomes interesting to see, to which extend these structural information disappear during diagenesis. The diagenetic pathway of steroids is well investigated but complex (see Fig. 3.24).

A major route leads to the so-called steranes, which represent the pure aliphatic hydrocarbon skeletons of the corresponding biomolecules (according to Fig. 3.17).

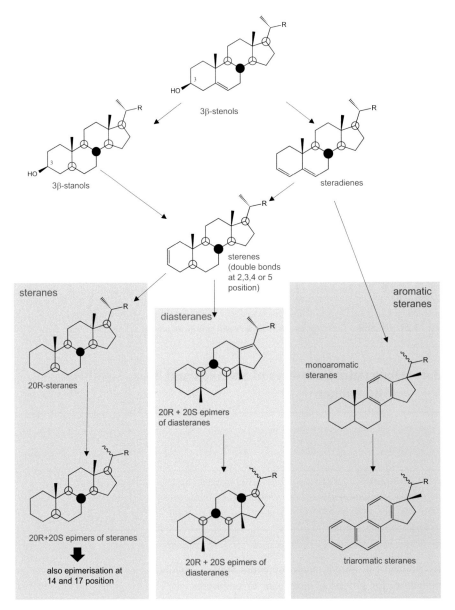

Fig. 3.24 General scheme of diagenetic pathways of steroids (modified and simplified after Mackenzie et al. 1982; Killops and Killops 2005; Peakmann and Maxwell 1988; van Kaam-Peters et al. 1998)

The diagenetic pathway comprises defunctionalization, shifts and following hydration of double bonds. As a last step certain epimerization at some chiral centers (e.g. C20) are observable. This pathway corresponds well with an already described diagenetic pathway of β-amyrin (see Sect. 3.2.2, Fig. 3.15)

An alternative route is evident under specific depositional conditions. At the stage of reduction of the sterenes a specific rearrangement can be observed forming finally the so-called diasteranes. This process results in an 'inverse' pattern of the methyl substituents at the alicyclic rings. Such rearrangement seems to be catalyzed by acidic conditions during diagenesis e.g. at acidic moieties in specific clay minerals. Hence, the appearance of diasteranes reflects specific sedimentary environments. The ratio of rearranged and regular steranes has been often used to differentiate argillaceous from carbonate source rocks.

Case Example: Clay minerals and steranes rearrangement

The influence of clay minerals on the formation of diasteranes has been intensively investigated eg. by van Kamm-Peters et al. (1998). Comparing the mineral composition of sedimentary rocks with the relative proportion of diasteranes vs regular steranes pointed to the clay mineral/TOC ratio as key parameter for diasterane formation.

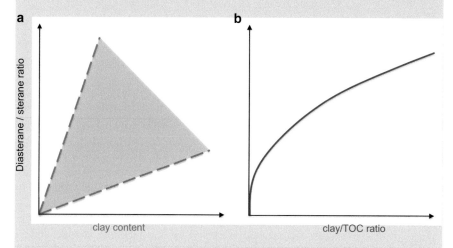

Figure: Schematic relationship between diasterane/sterane ratios and clay content (A) or clay/TOC ratio (B) (according to data adapted from van Kamm-Peters et al. (1998))

A completely different pathway is characterized by a sequential aromatization of the cyclohexyl moieties (see also Fig. 3.17). Significant intermediates are mono- and triaromatic compounds. In this case, the diagenetic products have been stabilized not by forming aliphatic hydrocarbons but by building up thermodynamically stable aromatic rings.

An overview on the major pathways of steroid diagenesis is given in Fig. 3.25. In this scheme the occurrence of diagenetic compounds (educts, intermediates and

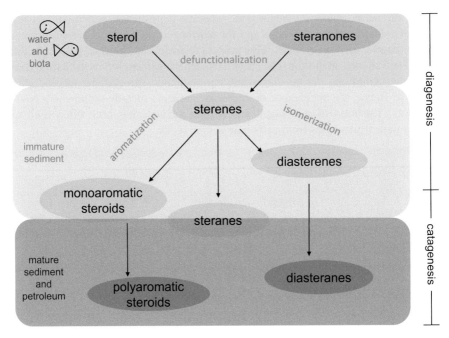

Fig. 3.25 Schematic attribution of general diagenetic conversion of steroids to principal reaction types and aquatic compartments (modified after Mackenzie et al. 1982)

final products) are linked to principal processes (e.g. defunctionalization, aromatization etc.) and the systems, where the individual reactions take place (e.g. water column, immature sediments etc.).

The last reactions steps in both main pathways, epimerization/isomerization as well as aromatization, are processes attributed to the beginning of catagenesis, where sediments enter the oil window. Hence, these chemical processes are used to follow the maturity during catagenesis by biomarker reflecting the stereochemical changes, e.g. the epimerization at C14, C17 or at C20. This is pointed out in Fig. 3.26, that illustrates the sequential change of the biogenic stereochemical configuration (α14, α17, 20R) to the thermodynamically more stable geochemical configuration (β14, β17, 20S). The intermediates reflect the grade of thermal maturity and, therefore, their relative proportions in fossil matter can be used as maturity indicator.

In summary, both diagenetic products from aromatization and epimerization are useful for tracing thermal maturity. In particular, sterane derived biomarker ratios are widely used as quantifying parameters for maturity studies of fossil matter. As already explained in Chap. 2, maturity parameters are based on the conversion of less stable biological molecules towards thermodynamically more stable chemofossils. Normally a ratio of products to educts is used as quantitative data

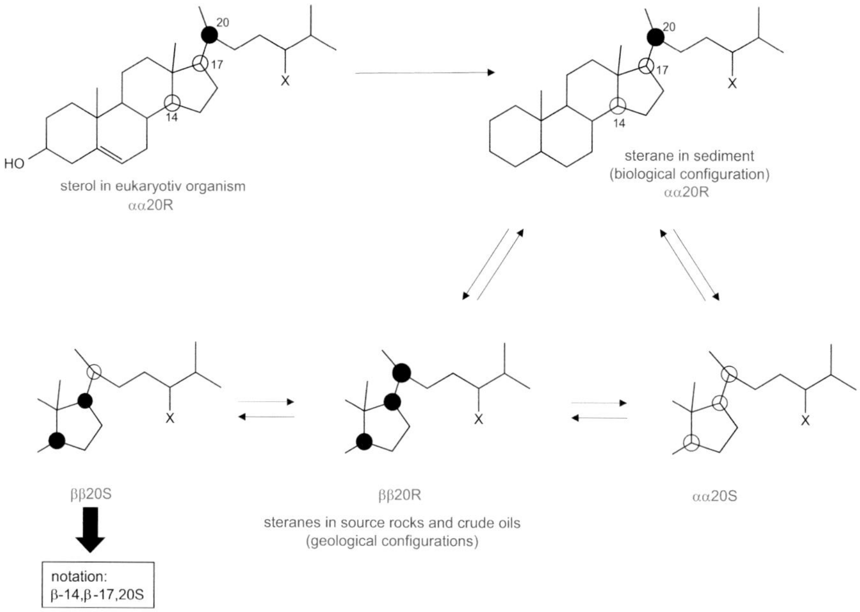

Fig. 3.26 Stereochemical changes of steroid compounds during dia- and catagenesis

starting with values of 0 for immature conditions. The underlying reactions can be either reversible or irreversible. Reversible conversions lead to steady state conditions between both, educts and products, represented by final biomarker ratios around 0.5. Irreversible processes result in the exclusive formation of the product over time and the corresponding final biomarker ratio of 1. How these parameters work is illustrated in Fig. 3.27 for some exemplary sterane biomarker ratios.

Noteworthy, endpoints of these conversions are related to different grades of maturity (classically expressed by vitrinite reflectance R_0, a microscopical parameters), hence each biomarker ratio works only in a selected region of thermal maturity. A correlation of sterane ratios to ranges of thermal maturity, in which they can be used as maturity parameter, are illustrated in Fig. 3.28.

Beside its application as maturity parameter, steranes provide information about the quality of organic matter as well as the depositional environment. As already mentioned, biogenic sterines can be indicative for their biogenic origin. However, during diagenesis those molecular properties differentiating the individual sterines (double bonds, chiral configuration etc.) have been lost. Solely the basic ring system including its methyl substituents and the side chain as well as the number of corresponding carbon atoms remains unaffected and can be used to obtain some information about the original biogenic composition of the organic matter. This rough characterization is widely applied by using triangular plots reflecting the

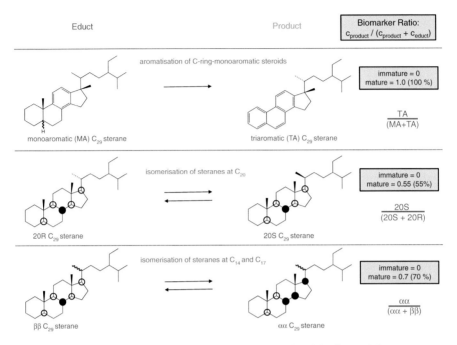

Fig. 3.27 Sterane biomarker ratios quantifying thermal maturity of fossil material

individual proportion of C_{27}-, C_{28}- and C_{29}-steroids. Different relative proportions reflect the original biogenic organic matter and, consequently, to some extend the depositional environment. The principal way of interpretation is illustrated in Fig. 3.29.

> ### *General Note*
>
> Steroid derived chemofossils have a wide range of biomarker functions. They characterize the original organic matter composition, and the depositional environment but act also as indicator for thermal maturity.

As pointed out, the usefulness of steranes for characterizing the original organic matter is restricted obviously to very general conclusions. One exception represents the dinosteroids. Their unusual methylsubstitution at C4 position remains also unaltered after diagenesis as illustrated in Fig. 3.30. Hence, the detection of dinosteranes in fossil samples gives a clear indication for the contribution of dinoflagellates to the original organic matter.

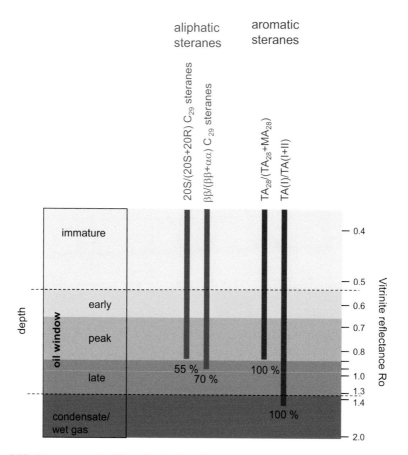

Fig. 3.28 Sterane maturity biomarker (ratios given as %) and corresponding ranges of validity given in vitrinite reflectance values or maturity levels with respect to oil generation (compiled and simplified after Peters et al. 2005; Mackenzie et al. 1982)

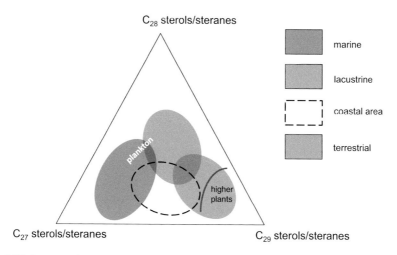

Fig. 3.29 Interpretation scheme for relative sterine/sterane distributions (represented as relative amounts of C_{27}-, C_{28}- and C_{29}-components) and its relation to potential source organism and corresponding depositional environments (compiled and simplified after Huang and Meinsheim 1979; Shanmugam 1985; Killops and Killops 2005)

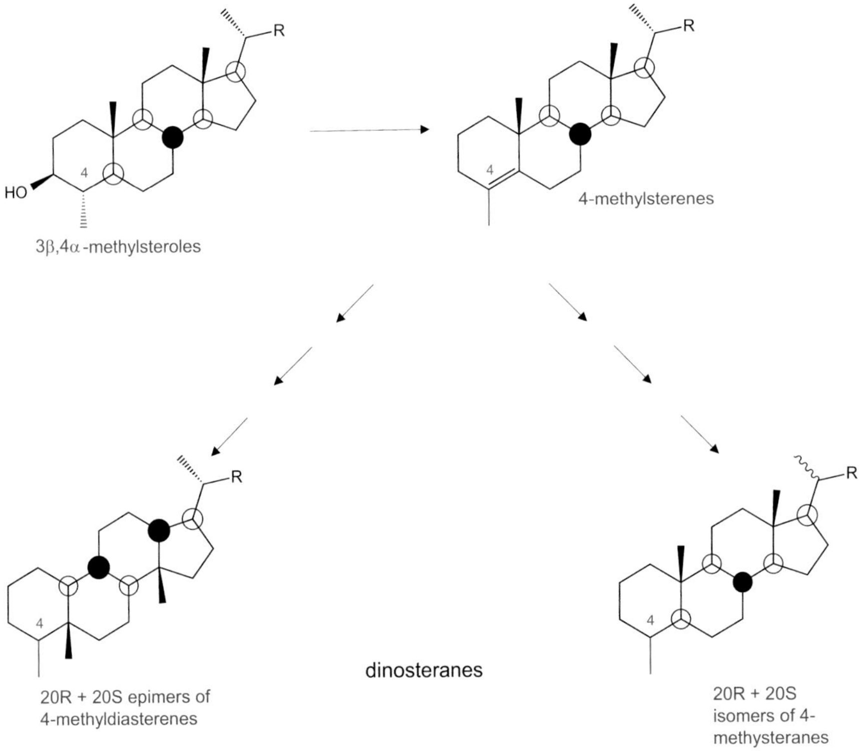

Fig. 3.30 Simplified diagenetic pathway of dinosterols (simplified and modified after Killops and Killops 2005; Peakman and Maxwell 1988)

Excursus: *Do homologue ratios of biomarkers remain constant throughout the diagenetic pathway?*

However, the question arises whether biogenic sterol distributions remain unaltered under diagenetic conditions. Or with other words, do the steranes in fossil matter reflect unambiguously the original biological composition? To verify the assumption that sterol composition persist unaltered in sediments over time, steroid alteration has been followed for the early stage of diagenesis by Gaskell and Eglinton (1976). As illustrated in the figure below, a study on steroid ratios in a lake sediment (30 cm) pointed clearly to unaltered relation of biogenic sterols and its first diagenesis products, the steranes. Both substance classes reflect the same changes with sediment depth, in particular between layer A and B. Hence, it can be assumed, that fossil sterane distributions reflect he original biological patterns

(continued)

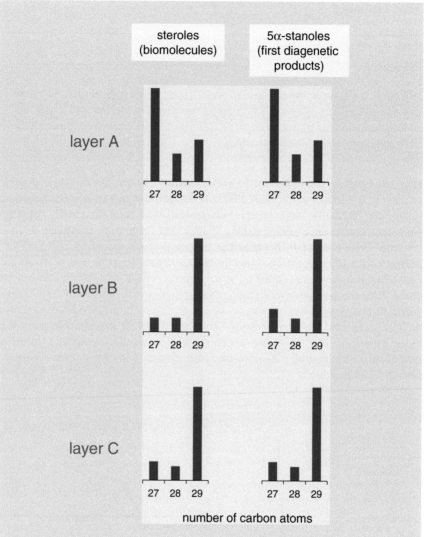

Figure: Sterol/stanol ratios in a lake sediment core (adapted and simplified from Meyers and Ishiwatari 1993)

3.5   Hopanes

Outlook

Lastly, hopanes represent an important group of biomarker with a unique analytical history. However, similarities to steroids with respect to biological function, molecular properties as well as diagenetic fate are obvious. Therefore, hopanes act also as biomarker e.g. for determining thermal maturity.

Hopanes represent one of the best known biomarker classes. The history of organic-geochemical research on hopanes differs as compared to nearly all other biomarker substances. Normally, biomolecules in organisms have been chemically characterized long before their corresponding biomarkers have been identified in fossil material. With respect to hopanes the chemical characterization appeared in a contrary order. Firstly, the so-called geohopanes, pure alicyclic hydrocarbons (see Fig. 3.31), have been described to be substantial constituents of oil, kerogen and coals. They are detectable as a group of homologues with side chain lengths up to 7 or 8 carbon atoms. The 'prototype' of the geohopanes, simply called hopane, is the C_{30} compound with an *iso*-propyl substituent at C21 positon. Members with lower carbon number are called 'norhopanes'. The smallest member exhibits no side chain and is named trisnorhopane. Homologues with longer side chains are named 'homohopanes'.

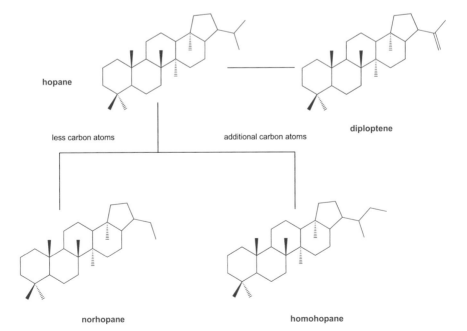

Fig. 3.31 Molecular structures of some geohopanes

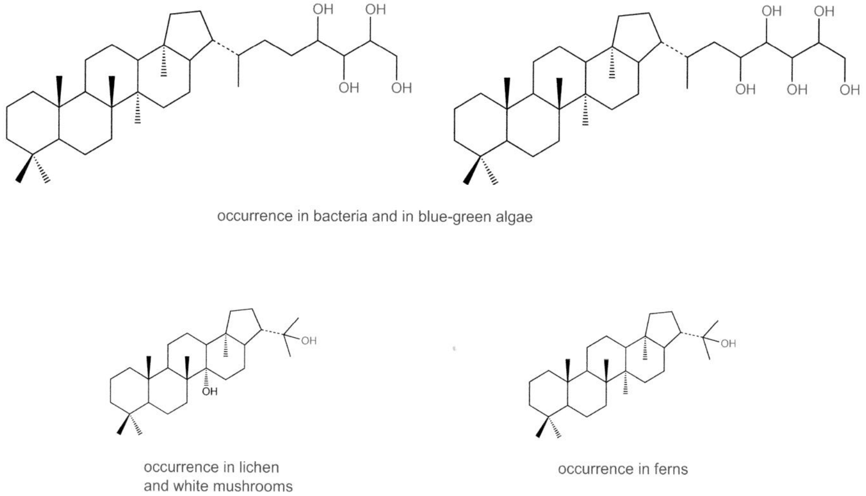

occurrence in bacteria and in blue-green algae

occurrence in lichen
and white mushrooms

occurrence in ferns

Fig. 3.32 Molecular structures of some biohopanes

Although their occurrence is abundant in fossil samples, the corresponding
biomolecules have not been known for a longer time. In particular, the source
organism for hopanes remained unknown and it seemed that an important pool of
biomolecules have been overseen for decades or even a century. However, years
after the structural elucidation of geohopanes, functionalized derivatives have been
detected to a minor extent in more simple plant species like ferns or lichens, but
finally to a high extent in bacteria (see Fig. 3.32). Following, the structural diversity
of biohopanes became visible by many re-analysis of bacterial matter. These
biohopanes exhibit the same carbon atom skeleton as compared to the geohopanes
but exhibit numerous hydroxy groups at the acyclic side chain.

Excursus: *How to become analytically invisible*

Biohopanes are a nice example for so-called amphoteric molecules. Such
molecules combine a large lipophilic moiety of aliphatic structure (cyclic or
acyclic) with a smaller functionalized part (e.g. substituted with hydroxy or
carboxylic groups) representing very hydrophilic properties. Amphoteric
molecules can act as mixing promoter for lipophilic and hydrophilic phases
(e.g. water and organic solvents). Therefore, detergents are typical representatives of amphoteric substances.

Analytical treatment of biological samples includes extraction as an
important step. During this extraction procedure organic analytes underlie a
partition between two phases mainly consisting of water (polar) and organic
nonpolar solvent. In this system amphoteric substances act similar as detergents and generate a mixed phase with micelles or a type of suspension at the

(continued)

interface of both main phases. Traditionally, this suspended mixed phase has been discharged by the analysts to work further with a pure organic layer. But together with this mixed phase also huge amounts of the amphoteric substances have been discarded. Another fraction of amphoteric substances remain on the particulate matter (e.g. cell residues) and become also ignored due to their analytical unavailability. Both has happened also with the biohopanes that have not identified earlier since they have been often poured away (as constituents of the 'mixed' phase or together with the extracted particulate residues) during the analytical procedure.

Figure: Generation of a suspension interface (*red dotted line*) between water and organic solvent during liquid/liquid extraction in a separatory funnel

From a structural point of view hopanes exhibit a high similarity with steroids. Instead of three condensed six-membered rings and one five-membered ring the hopanes are built of four condensed six-membered rings and one five-membered ring. The individual rings are condensed with alternating linear and angular arrangement. However, hopanes exhibit a higher number of methyl substituents and chiral carbon atoms.

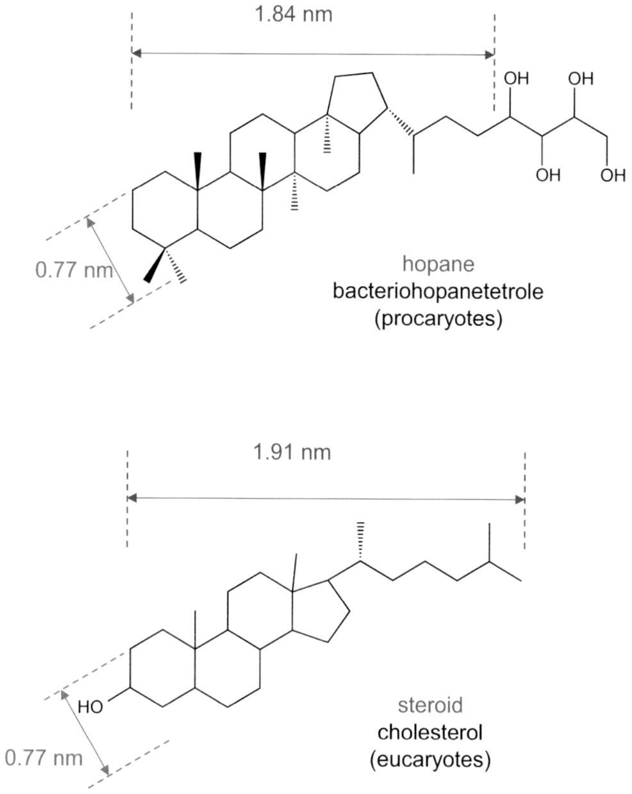

Fig. 3.33 Comparison of length and width of sterols and biohopanes (according to Ourisson 1986)

A high structural similarity is also reflected by the overall dimensions of the molecules as depicted in Fig. 3.33. The overall length but in particular the 'shoulder length' (0.77 nm) are nearly the same. Beside the similar shape also amphoteric properties are obvious for both substance classes (see excursus).

These similarities are based on the biosynthesis pathway. The biotic synthesis of both substance groups uses squalene as educt, but the way of cyclisation differs. On the one hand, synthesis of hopanes does not need necessarily an epoxidation as initial reaction step and the total number of cyclization (five rings) is higher (see Fig. 3.34).

Similar size and similar amphoteric properties might point to a related biological function. And indeed, in the same way as steroids act as membrane constituents, in particular to harden the cell walls, hopanes adopt this function in bacteria cells. They are inserted into the phosphor lipid based double layer membranes as delineated in Fig. 3.35. Consequently, hopanes can act as perfect indicators for bacteria related contributions to fossil organic matter.

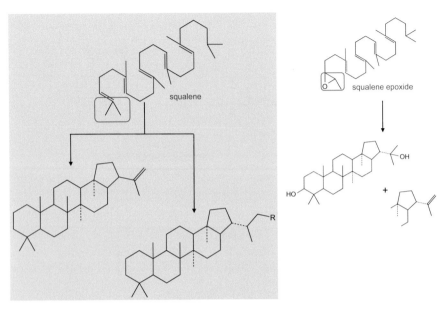

Fig. 3.34 Scheme of biosynthesis pathway of hopanes in comparison to steroids (simplified after Ourisson et al. 1979; Ourisson 1986)

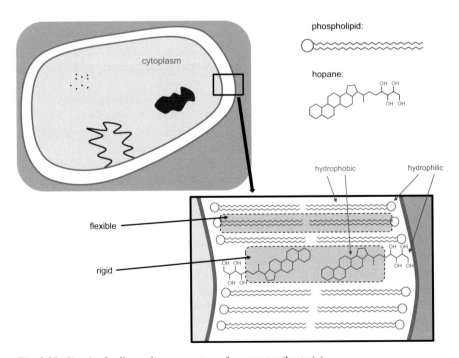

Fig. 3.35 Sketch of cell membrane structure of eucaryotes (bacteria)

3.6   Diagenetic Fate of Hopanes

The diagenetic fate of hopanes consists of three main transformations, the loss of hydroxy groups, a shortening of the side chain and, lastly, the epimerization or racemization at certain chiral carbon atoms. The defunctionalization is realized either as simple reduction of the hydroxy groups or by oxidation to carboxylic groups and further decarboxylation, which results in a side chain shortening (see Fig. 3.36). The competition of both defunctionalization routes is controlled by the oxygen availability during deposition and sedimentation. However, both types of reaction seem to occur to various extents in parallel or sequentially and result finally in the formation of the homologues series.

With ongoing diagenesis and catagenesis the stereochemical properties become modified at distinct chiral centers. One modification converts the biologically determined R-configuration at the first side chain carbon atom (C22) of all homohopanes to a more or less racemic composition of S- and R-isomers (see Fig. 3.37). The $R/(R+S)$ ratio is used as marker for thermal maturity in the range from immaturity to an early stage maturity (see Fig. 3.38). The final value of 0.6 represents a steady state with an isomeric composition of 60 % of S-enantiomer and 40 % of R-enantiomer reflecting a slightly higher thermodynamical stability of the S-enantiomer.

Another important change of stereochemical properties is located in the five-membered ring at C17 and C21. The stereo configuration at this carbon atoms is biologically preassigned as 17β, 21β-configuration (Fig. 3.39).

During diagenesis this configuration changes to a final product with 17α, 21-β-configuration. However, temporarily also the 17β, 21β-configuration appears during the diagenetic process. The conversion from β, β- to α, β- partially via β,α-isomers is based on the relative thermodynamical stabilities of the diastereomers as illustrated in Fig. 3.40. The lower energy level of the α, β-isomer (also called moretanes) triggers the diagenetic reaction and allows to use the relative proportions of the biological and the thermodynamical more stable isomer to quantify thermal maturity. The corresponding biomarker ratio is valid as maturity parameter roughly in the same range as compared to the 22R/22S-biomarker (see Fig. 3.38). Further on, the energetic point of view also explains the sporadically appearance of β, α-isomer as intermediates along the conversion pathway. Further on, it becomes clear why α, α-isomers as the most unstable diastereomers have not been identified in sedimentary systems.

Beside stereochemical conversions also rearrangements are observed during hopane diagenesis. The most prominent example is the shift of a methyl group in trisnorhopane from the C17 (thermodynamical less stable, named T_m) to the C18 position (thermodynamical more stable, named T_s) as illustrated in Fig. 3.37. Since also this reaction is initiated by thermal stress, the corresponding biomarker ratio $T_s/(T_s + T_m)$ is used as common maturity parameter with a validity ranging to the late maturity (see Fig. 3.38).

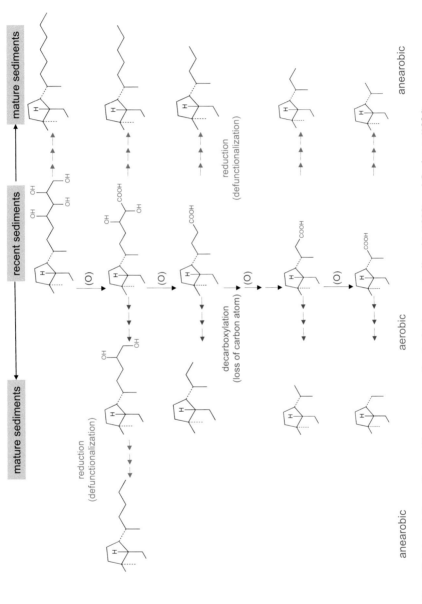

Fig. 3.36 Simplified degradation pathway of hopanes according to Ourisson and Albrecht (1992) and Ourisson (1986)

| Educt | → | Product | Biomarker Ratio: $c_{product} / (c_{product} + c_{educt})$ |

isomerisation of hopanes at C_{22}

22R C_{32} hopanes → 22S C_{32} hopanes

immature = 0
mature = 0.6 (60 %)

$$\frac{22S}{(22S + 22R)}$$

Tm (less stable) → Ts (more stable)

immature = 0
mature = 1.0 (100%)

$$\frac{T_s}{(T_s + T_m)}$$

notation:
structures of 17α-22,29,30-trisnorhopane (Tm)
and   18α-22,29,30-trisnorneohopane (Ts)

Fig. 3.37 Hopane based biomarker ratios determining thermal maturity

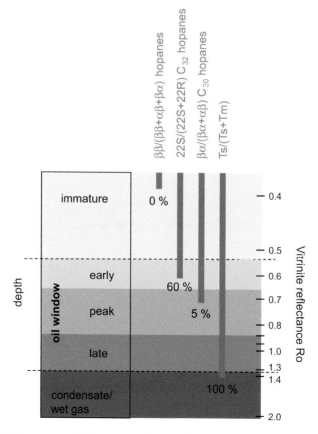

Fig. 3.38 Validity range in which exemplary hopane biomarker work well (modified after Peters et al. 2005; Mackenzie 1984)

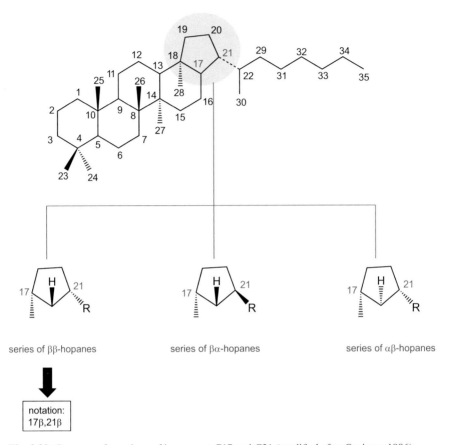

Fig. 3.39 Stereo configurations of hopanes at C17 and C21 (modified after Ourisson 1986)

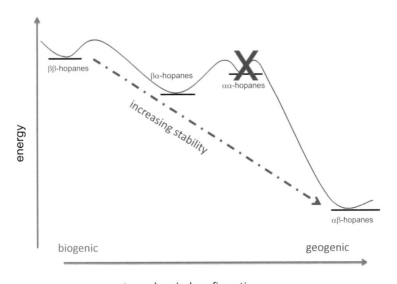

Fig. 3.40 The thermodynamical grading of hopane C17/C21-diastereomers

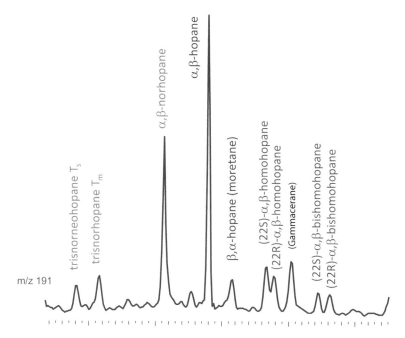

Fig. 3.41 Gas chromatogram of hopanes in a source rock extract

In summary, the different degradation pathways and the stereochemical conversions produce a complex mixture of geohopanes consisting of homologues series and various stereoisomers. This mixture can be analytically resolved by gas chromatography and an example of such corresponding gas chromatogram of hopanes is presented in Fig. 3.41.

General Note

The basic importance of hopane derived biomarkers is related to their function as thermal maturity parameters.

Excursus: *Hopanes as marker for petrogenic emissions*

Since hopanes can represent fossil organic matter the occurrence of hopane patterns in recent ecosystems can point to petrogenic pollution. Mainly in addition to other petrogenic indicators (e.g. sterane ratios, unresolved complex mixture UCM, CPIs etc.), the hopane biomarker are useful to identify

(continued)

mature matter in recent sediments. Such observations are solely explainable by petrogenic contaminations from fossil products. An example of hopanes detected from surface sediments of the Jakarta Bay is given below. The calculated biomarker ratios point to a maturity within the oil window.

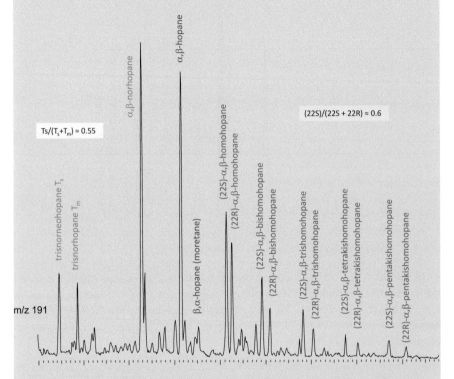

Figure: Gas chromatogram of hopanes in a surface sediment of Jakarta Bay contaminated with petrogenic emissions. Here biological and fossil hopane signatures are superimposed

3.7 Glycerol Dialkyl Glycerol Tetratether Lipids (GDGTs) and Their Organic-geochemical Relevance

Outlook

GDGTs represent a biomarker group with high potential to reconstruct paleoenvironmental conditions. The systematic relationship between their structural properties and modifications with changes of environmental conditions is described.

Isoprenoid moieties are also substantial components in glycerol dialkyl glycerol tetratether lipids, due to their long name better known as GDGTs. These compounds are important membrane constituents in archaea, procaryotes widespread distributed in marine and terrestrial environments. They consist typically of two glycerine moieties connected via ether bonded long chain alcohols of isoprenoic structure. Formally, these alcohols can be considered as covalently linked biphytols. Both hydroxy groups at the end of these tetraterpenediols are linked via ether bonds to the two glycerine molecules. In sum, two of the three hydroxy groups of glycerine are linked but the third group remains for connection to polar head groups (e.g. phosphate) in intact cells. These head groups get rapidly lost after release to the geosphere.

The spectrum of GDGTs is obtained by slight structural changes in the linking isoprenoic moieties. Dominantly cyclization forming five- and to a very low extent six-membered rings is observed as illustrated in Fig. 3.42. The individual derivatives are named by a special notation by using the abbreviation GDGT followed by a running Arabic number.

The chemical structure of GDGTs is characterized by two more polar sectors at the ends and a lipophilic middle block. This feature is used by archaea to form monolayer cell membranes (see Fig. 3.43), whereas bacteria and eukaryotes produce bilayer cell membranes dominantly with phospholipids and related biomolecules (as described in more detail in Sect. 4.3).

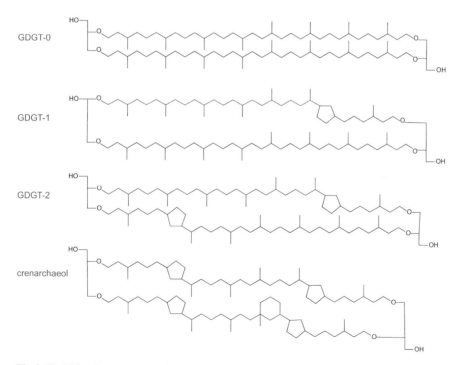

Fig 3.42 Molecular structure and notation of isoprenoid DGDTs (isoGDGTs)

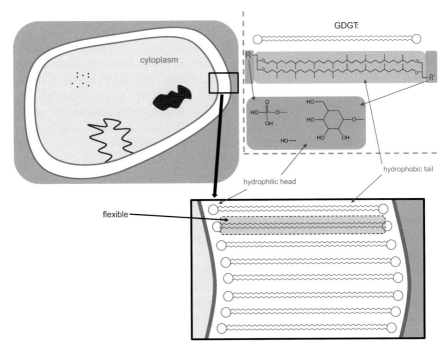

Fig. 3.43 GDGTs in cell membranes of archaea

Beside isoprenoid GDGTs also branched GDGTs have been identified, which exhibit alkyl chains with less methyl substituents (e.g. 4–6, see Fig. 3.44). These membrane lipids are assumed to appear dominantly in terrestrial archaea and bacteria. Their notation is similar to those of isoprenoid GDGTs, but Latin numbers are used instead of Arabic ones (see Fig. 3.44).

However, the usefulness of GDGTs with respect to Organic Geochemistry is not limited to their potential to indicate the contribution of archaea or related species to the organic matter. The GDGTs pattern differs under varying environmental conditions. This accounts for both isoprenoid and branched GDGTs. These are very well examples to point out the relation of structural properties or changes and their organic-geochemical implications.

Laboratory observations that the degree of cyclization of isoGDGTs in cultivated microbes correlates well with the water temperature opened the possibility to use this parameter to estimate the paleo sea-surface temperature from sedimentary GDGTs. A corresponding parameter well established in Organic Geochemistry is the TEX86 index. Noteworthy, since the quantitative correlation between temperature and degree of cyclization is *ab initio* not predictable, the TEX86 needs to be calibrated empirically, which is a sensitive task. Therefore, several different calibration studies using diverse sedimentary systems are reported. An exemplifying approximation according to Schouten et al. (2002) is given in Fig. 3.45.

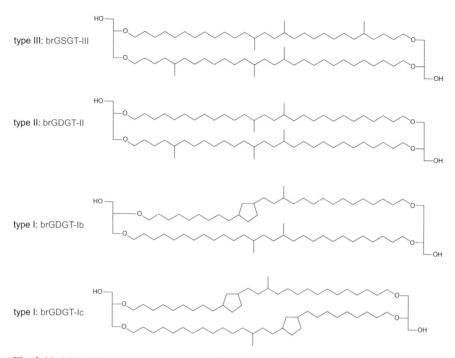

Fig. 3.44 Molecular structure and notation of branched GDGTs (brGDGTs)

$$TEX_{86} = \frac{[GDGT-2]+[GDGT-3]+[cren']}{[GDGT-1]+[GDGT-2]+[GDGT-3]+[cren']}$$

$$TEX_{86} = 0.015 \times SST + 0.27 \quad (r^2 = 0.92, n = 43)$$

TEX_{86}

Fig. 3.45 The TEX86 parameter (Schouten et al. 2002)

Accordingly, also for branched GDGTs a paleothermometer approach exist using the co-called MBT and CBT parameters. This approach is a little bit more complex as compared to TEX86 and is based on two key observations: (i) degree of cyclization of brGDGTs corresponds well with soil pH and (ii) the degree of methylation as well as of cyclization of brGDGTs correlate well with soil temperature or to the related parameter mean ambient air temperature (MAAT), respectively. Hence, the corresponding parameters CBT (cyclization of branched GDGTS) and MBT (methylation of branched GDGTS) can be used to obtain information about pH and paleotemperatures in terrestrial systems like soils and lakes (see Fig. 3.46). However, also these parameters need empirical calibration, which is a critical issue.

$$CBT = -\log\left(\frac{[brGDGT - Ib] + [brGDGT - IIb]}{[brGDGT - I + brGDGT - II]}\right)$$

$$CBT = 3.33 - 0.38 \times pH$$

CBT

$$MBT = \frac{[brGDGT - I] + [brGDGT - II] + [brGDGT - III]}{\sum[brGDGTs]}$$

$$MBT = 0.122 + 0.187 \times CBT + 0.020 \times MAAT$$

MBT

Fig. 3.46 CBT and MBT parameters and their usage for approximation of paleoconditions in terrestrial systems according to Thierny (2012)

General Note

GDGTs allow a quantitative relationship of biomarker composition and paleoenvironmental information. The usefulness of these GDGT based parameter is related not only to the empirical equations but also to the widespread occurrence of the biological sources, the archaea, allowing a broad application.

Case Example: GDGT as paleotemperature proxy

Both GDGT proxy have been used by Weijers et al. (2011) to reconstruct paleoenvironmental conditions during peatification. They tested MBT and CBT measured in different peat bogs for estimating e.g. mean annual air temperature. Finally, they interpreted temperature changes at stratigraphic boundaries with changing types of peat (e.g. dominance of *sphagnum* or *carex*). This would imply restrictions in the general application of GDGT proxy (Fig. 3.47).

(continued)

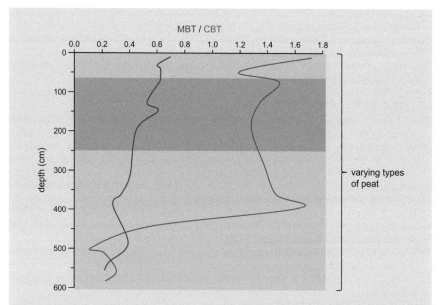

Fig. 3.47 The course of MBT and CBT values in a representative peat bog. Note the shifts in the CBT profile at boundaries representing the transition between different types of peat (adopted and simplified from Weijers et al. 2011)

References

Gaskell SJ, Eglinton G (1976) Sterols of a contemporary lacustrine sediment. Geochim Cosmochim Acta 40:1221–1228

Hautevelle Y, Michels R, Malartre F, Trouiller A (2006) Vascular plant biomarkers as proxies for palaeoflora and palaeoclimatic changes at the Dogger/Malm transition of the Paris Basin (France). Org Geochem 37:610–625

Huang WY, Meinschein WG (1979) Sterols as ecological indicators. Geochim Cosmochim Acta 43:739–745

Killops S, Killops V (2005) Introduction to organic geochemistry, 2nd edn. Blackwell Publishing, Oxford

MacKenzie AS, Brassell SC, Eglinton G, Maxwell JR (1982) Chemical fossils: the geological fate of steroids. Science 217:491–504

MacKenzie AS (1984) Application of biological markers in petroleum geochemistry. In: Brooks J, Welte DH (eds) Advances in petroleum geochemistry, vol 1. Academic, London, pp 115–214

McGovern PE, Mirzoian A, Hall GR (2009) Ancient Egyptian herbal wines. Proc Natl Acad Sci U S A 106:7361–7366

Meyers PA, Ishiwatari R (1993) The early diagenesis of organic matter in lacustrine sediments. In: Engel MH, Macko SA (eds) Organic geochemistry – principles and applications. Plenum Press, New York/London, pp 185–209

Moldowan JM (1994) The molecular fossil record of oleanane and its relation to angiosperms. Science 265:768–771

Otto A, Simoneit BRT (2002) Biomarkers of holocene buried conifer logs from Bella Coola and north Vancouver, British Columbia, Canada. Org Geochem 33:1241–1251

Ourisson G (1986) Vom Erdöl zur Evolution der Biomembranen. Nachrichten aus Chemie, Technik und Laboratorium 34:8–14

Ourisson G, Albrecht P, Rohmer M (1979) The hopanoids – palaeochemistry and biochemistry of a group of natural products. Pure Appl Chem 51:709–729

Ourisson G, Albrecht P (1992) Hopanoids 1. Geohopanoids: the most abundant natural products on earth? Acc Chem Res 25:398–402

Peakman TM, Maxwell JR (1988) Early diagenesis pathways of steroid alkenes. Org Geochem 13:583–592

Peters KE, Walters CC, Moldowan JM (2005) The biomarker guide, 2nd edn. Cambridge University Press, Cambridge

Schouten S, Hopmans EC, Schefuß E, Sinninghe Damste JS (2002) Distributional variations in marine crenachaeotal membrane lipids: a new tool for reconstructing ancient sea water temperatures? Earth Planet Sci Lett 204:265–274

Shanmugam G (1985) Significance of coniferous rain forests and related organic matter in generating commercial quantities of oil, Gippsland Basin, Australia. Bull Am Assoc Pet Geol 69:1241–1254

Summons RE, Bradley AS, Jahnke LL, Waldbauer JR (2006) Steroids, triterpenoids and molecular oxygen. Philos Trans R Soc B 361:951–968

Thierny JE (2012) GDGT thermometry: lipid tools for reconstructing paleotemperatures. In: Reconstructing earth's deep-time climate. Paleontol Soc Pap 18:115–131

Van Aarsen BGK, Alexander R, Kagi RI (2000) Higher land plant biomarkers reflect palaeovegetation changes during Jurassic times. Geochim Cosmochim Acta 64:1417–1424

Van Kaam-Peters HME, Köster J, Van Der Gaast SJ, Dekker M, De Leeuw J, Sinninghe-Damste JS (1998) The effect of clay minerals on diasterane/sterane ratios. Geochim Cosmochim Acta 62:2923–2929

Volkman JK (2005) Sterols and other triterpenoids: source specificity and evolution of biosynthetic pathways. Org Geochem 36:139–159

Weijers JWH, Steinmann P, Hopmans EC, Schouten S, Sinninghe Damste JS (2011) Bacterial tetraether membrane lipids in peat and coal: Testing the MBT-CBT temperature proxy for climate reconstruction. Org Geochem 42:477–486

Further Reading

Ourisson G, Albrecht P (1992) Hopanoids 2. Biohopanoids: a novel class of bacterial lipids. Acc Chem Res 25:403–408

Summons RE, Bradley AS, Jahnke LL, Waldbauer JR (2006) Steroids, triterpenoids and molecular oxygen. Philos Trans R Soc B 361:951–968

Schouten S, Hopmans EC, Sinninghe Damste JS (2013) The organic geochemistry of glycerol dually glycerol tetraether lipids: a review. Org Geochem 54:19–61

Volkman JK (2005) Sterols and other triterpenoids: source specificity and evolution of biosynthetic pathways. Org Geochem 36:139–159

Chapter 4
Polyketides

A second important substance class in Organic Geochemistry are polyketides. Many so-called lipids, which comprise in principal natural substances with elevated lipophilicity, belongs to this substance class or exhibit polyketides as major part of their molecules. Biochemical important examples are fatty acids, fatty oils, waxes, phospholipids and further membrane lipids. These compounds play important roles in building up cells, storage of energy in food cycles, as protecting agents in protective films and further basic physiological functions.

Polyketides are defined by their principal biosynthesis pathway. A scheme of the initial steps and the total synthesis cycles are given in Fig. 4.1. Starting with two common biosynthetically activated precursors, the acetyl-CoA (already introduced by the biochemistry of isoprenoids) and the malonyl-CoA, an addition by condensation and following a removal of the hydroxy group (by dehydration and reduction) lead to a C4 analogue of acetyl-CoA, the butyryl-CoA. These two steps, addition and defunctionalization, are repeated until the desired final length of the biomolecule is achieved. In summary, the overall synthesis is simply the sequential addition of the C2 units. As a result, polyketides are generally characterized by unbranched and even-numbered carbon chains. These are specific structural properties which allow to identify polyketides easily among the huge pool of biomolecules. Following, the most important polyketides, their structural properties and their diagenetic fate are described in more detail.

General Note

Polyketides are easily recognizable by their unbranched long carbon chains. This structural property is a distinctive feature in comparison to the regularly branched isoprenoids.

© Springer International Publishing Switzerland 2016
J. Schwarzbauer, B. Jovančićević, *From Biomolecules to Chemofossils*,
Fundamentals in Organic Geochemistry, DOI 10.1007/978-3-319-25075-5_4

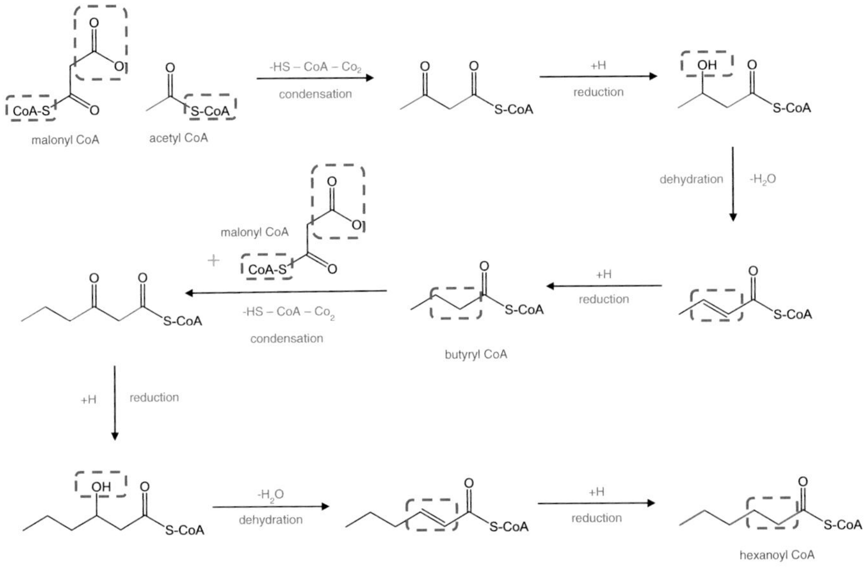

Fig. 4.1 Simplified biosynthesis pathway for polyketides

4.1 Fatty Acids

Outlook

Fatty acids represent the most important representatives of polyketides. Their principal structural diversity is discussed here.

Fatty acids are simply long chain alkanoic acids with chain length typically above 10 carbon atoms and normally with unbranched chains. The long aliphatic chain determines the lipophilic properties of these molecules, whereas the carboxylic group exhibits polar properties. Biological most important fatty acids are the hexadecanoic and octadecanoic acids, also named palmitic and stearic acids. Since fatty acids have been investigated for a very long time period, they are usually named unsystematically with trivial names. Further on, to avoid the extended usage of long names a common shorthand notation has been established for fatty acids. This notation just specifies the chain length and, separated by a colon, the number of double bonds. A correlation of some very important fatty acids, their trivial names as well as their shorthand notation are given in Table 4.1.

Fatty acids are divided into several sub classes according to their structural variations. Firstly, saturated fatty acids can be distinguished from unsaturated derivatives, which exhibit at least one double bond (see Table 4.2). Noteworthy, many multiple unsaturated acids are known. Most prominent unsaturated fatty acids

Table 4.1 Systematic and trivial names as well as the corresponding shorthand notation of C_{10} to C_{28} fatty acids

Systematic name	Trivial name	Shorthand notification
Decanoic acid	Capric acid	10:0
Dodecanoic acid	Lauric acid	12:0
Tetradecanoic acid	Myristic acid	14:0
Hexadecanoic acid	Palmitic acid	16:0
Heptadecanoic acid	Margaric acid	17:0
Octadecanoic acid	Stearic acid	18:0
Icosanoic acid	Arachidic acid	20:0
Docosanoic acid	Behenic acid	22:0
Tetracosanoic acid	Lignoceric acid	24:0
Hexacosanoic acid	Cerotic acid	26:0
Octacosanoic acid	Montanic acid	28:0

are the C_{18} acids with one, two or three double bonds – oleic, linoleic and linolenic acids. Unsaturated fatty acids exhibit principally a *cis*-configuration at the double bonds as exemplified in Table 4.2.

For unsaturated fatty acids two different systems to allocate the positions of the double bonds exist. Firstly, the systematic numeration starts at the carboxylic group which leads to a systematic designation that results e.g. for linoleic acid in the positions 9 and 12 (exemplified in Fig. 4.2). The corresponding entire name is *cis, cis*-9,12-octadecadienoic acid. An alternative nomenclature starts the numeration at the non-functionalized end of the carbon chain. This numeration is indicated by the letter ω (omega as last letter in the Greece alphabet indicates generally the end position). Correspondingly, linoleic acid is alternatively named ω(6,9)-octadecadienoic acid. Indication of *cis*-configuration is not necessary here, since only biogenic fatty acids are named according to the ω-nomenclature and this implies *cis* double bonds. The reason for the second nomenclature is related to a similar biochemical or physiological relevance of unsaturated fatty acids with double bond positions with the same distance to the carbon chain end. A representative example are the so-called ω3-fatty acids (see Fig. 4.2) playing an essential role in human nutrition.

These two different numerations are also considered by the shorthand notation for unsaturated fatty acids. At first, the number of double bonds is given by the number behind the colon followed by the positions and the abbreviated configuration (c for *cis* or t for *trans*) written in brackets. Alternatively, the configuration can be given as prefix in full length. If the ω-notation is used, the prefix ω is added to the positions as exemplified for some unsaturated fatty acids in Table 4.3.

Table 4.2 Compilation of exemplary saturated and unsaturated fatty acids

	Trivial Name	Formula	Structure	Source
Saturated	Lauric acid	$C_{11}H_{23}CO_2H$		Laurel oil, palm oil, animal fats
	Myristic acid	$C_{13}H_{25}CO_2H$		Coconut oil, palm oil, animal fats
	Palmitic acid	$C_{15}H_{31}CO_2H$		Palm oil, animal and vegetable fats, beeswax
	Stearic acid	$C_{17}H_{35}C_{02}H$		Animal and vegetable fats
Unsaturated	Oleic acid	$C_{17}H_{33}CO_2H$		Corn oil, cottonseed oil
	Linoleic acid	$C_{17}H_{31}CO_2H$		Olive oil, fish oil, com oil, linseed oil
	Linolenic acid	$C_{17}H_{29}CO_2H$		Linseed oil
	Arachidonic acid	$C_{19}H_{31}CO_2H$		Sardine oil, com oil, animal fats

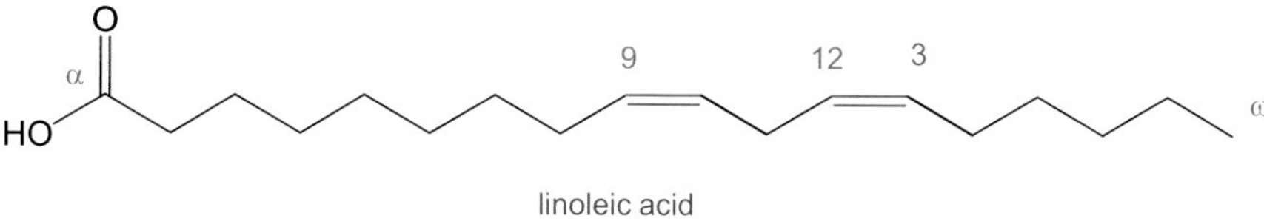

linoleic acid

Fig. 4.2 Numeration systems for unsaturated fatty acids

Table 4.3 Systematic and trivial names as well as the corresponding shorthand notation of some unsaturated fatty acids

Systematic name	Trivial name	Shorthand notification			
cis-hexan-9-enoic acid	Palmitoleic acid	cis-16:1(9)	16:1(9c)	(ω-7)-16:1	16:1(ω7)
cis,cis,octadec-9,12-dienoic acid	Linoleic acid	cis,cis-18:2 (9,12)	18:2(9c,12c)	(ω-6)-18:2	18:2(ω6)
trans,trans-octadeca-9,12-dienoic acid	Linelaidic acid	tran,trans-18:2 (9,12)	18:2(9t,12t)	(ω-6)-18:2	18:2(ω6)

Excursus: *ω3-fatty acids*

The relevance of the ω – notation is obvious for a special sub group of fatty acids, the ω3-fatty acids. These fatty acids (e.g. linolenic acid, eicosapentaenoic acid) are essential for humans, we cannot biosynthesize them by ourselves, but we need to take them up by food. Therefore, in former times ω3-fatty acids have been called *vitamin F*. Elevated concentrations can be found in fishes, algae and plants.

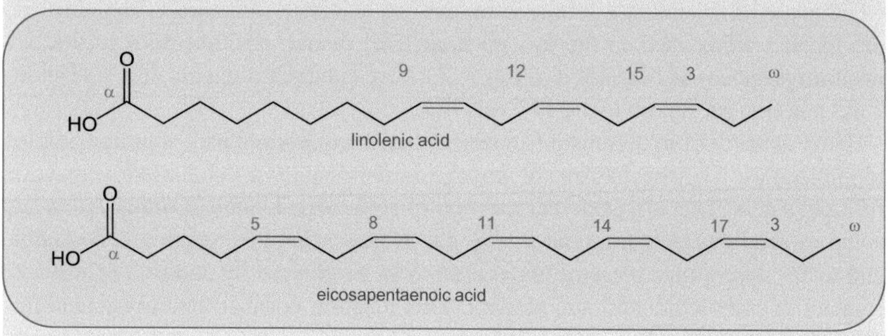

linolenic acid

eicosapentaenoic acid

A further sub class of fatty acids are the methyl branched derivatives. The unusual insertion of a branching is restricted to fatty acids biosynthesized by

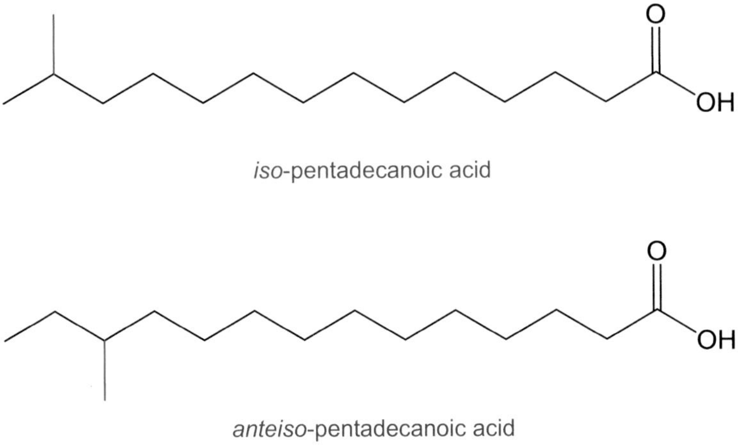

iso-pentadecanoic acid

anteiso-pentadecanoic acid

Fig. 4.3 Molecular structures of *iso*- and *anteiso*-pentadecanoic acid

Table 4.4 Example for the nomenclature of *iso*- and *anteiso* fatty acids

Total number of carbon atoms	Systematic name	Trivial name	Non specific	Iso/ anteiso	Specific
16	14-methylpenta-decanoic acid	Isopalmitic acid	br-16:0	i-16:0	14-Me-15:0
16	13-methylpenta-decanoic acid	Anteiopalmitic acid	br-16:0	a-16:0	13:Me-15:0
19	10-methylocta-decanoic acid	Tuberculostearic acid	br-19:0	–	10-Me-18:0

specific organism. Branched fatty acids derive dominantly from bacteria and, therefore, are indicative for bacterial organic matter. Only two different substitution isomers are biochemically synthesized, the *iso*- and *anteiso*-isomers. The positions are located either next to the last position (*iso*) or one position prior to the *iso*-position (*anteiso*) as illustrated in Fig. 4.3. Corresponding nomenclature and short-hand notation are summarized in Table 4.4.

Fatty acids exist in organism to a large extent as important part of more complex biomolecules. The mechanism of linkage is dominantly a condensation reaction forming esters. This is a principal reaction of acids and alcohols forming esters and water as products as figured out in Fig. 4.4. Noteworthy, this reaction is reversible and under appropriate circumstances esters can be cleaved by addition of water to re-react to carboxylic acid and alcohol. This reaction is called hydrolysis and this ester cleavage works best under alkaline conditions.

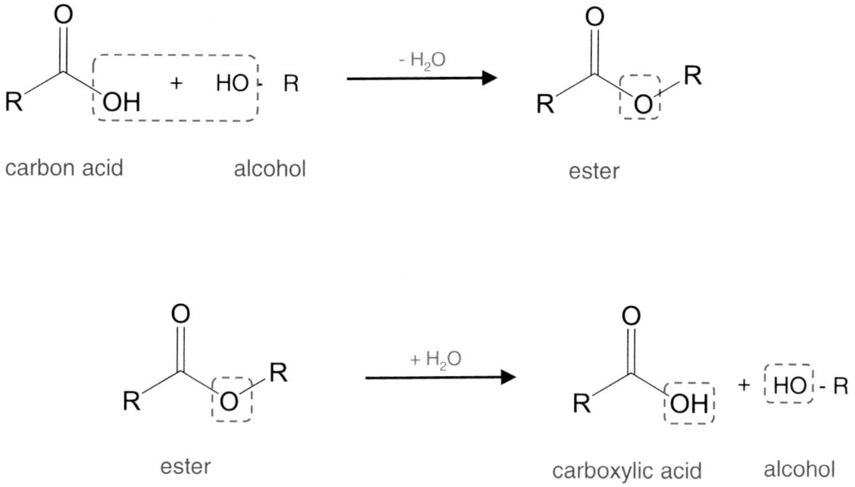

Fig. 4.4 Reaction scheme of esterification and the reversible reaction, the hydrolysis

4.2 Fats and Waxes

Outlook

Fatty acids are part of many important biological products such as fats and waxes. Their occurrence and their influence on physico-chemical properties are introduced.

Most important biomolecules with fatty acid moieties are fats. Fats act as one of the major biological energy storage but have certainly also further more complex functions in organisms. From a chemical point of view, fats are the triple esters of the alcohol glycerol, exhibiting three hydroxy groups that all can be linked via ester bonds forming so-called triglycerides. Fats as triglycerides can obtain up to three different but also just one or two different fatty acids. An example for the formation of an individual fat molecule is given in Fig. 4.5.

As exemplified in Fig. 4.5, saturated as well as unsaturated fatty acids can be incorporated. The proportion of unsaturated in relation to saturated moieties has a huge impact on some physico-chemical properties e.g. the melting points as discussed in the following. Insertion of double bonds in alkyl chains changes the steric conditions in particular for *cis* configurations that are common in natural fatty acids. The alkyl chain gets a fixed bend and, therefore, the molecules have a higher steric demand (see Fig. 4.6). Hence, the average distance between the molecules is higher by a higher degree of unsaturation. Following the intermolecular interactions (e.g. Van-der-Waals forces) become less intensive and, consequently, it needs less energy to transfer these molecules from the solid to a liquid state. This can be followed by distinguishing solid fats and liquid fatty oils by their relative proportions of unsaturated fatty acids as exemplified in Table 4.5

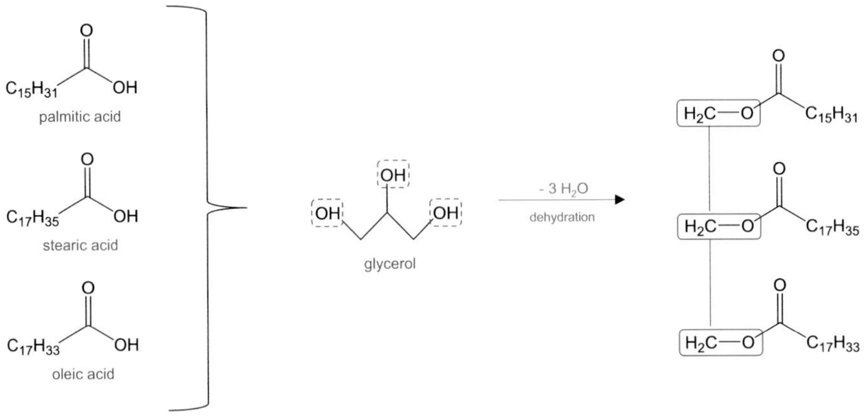

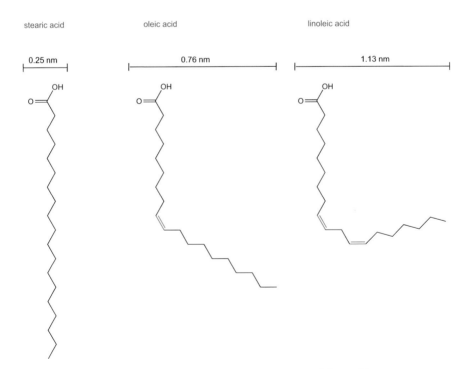

Fig. 4.5 Esterification of glycerol with three different fatty acids forming an individual fat molecule

Fig. 4.6 Comparison of steric demand of saturated and unsaturated fatty acids

Table 4.5 Melting points and corresponding state at room temperature for typical oils and fats and corresponding proportion of exemplary saturated and unsaturated fatty acids

	Oil	Melting range [°C]	Palmitic acid	Stearic acid	Oleic acid	Linoleic
Solid at room temperature	Coconut oil	20–22	4–10	1–5	2–10	0–2
	Palm oil	Mostly <25	45–54	3–6	38–40	5–11
	Butter	30	23–26	10–15	30–40	4–5
	Sebum	45	24–52	14–52	35–48	2–4
Liquid at room temperature	Castor oil	−18 to −10	0–1	–	0–9	5–7
	Olive oil	−5 to 9	5–15	1–4	69–84	4–12
	Rape oil	5–15	0–1	0–2	20–58	10–15
	Linsed oil	−26 to −16	4–7	2–5	9–58	5–45

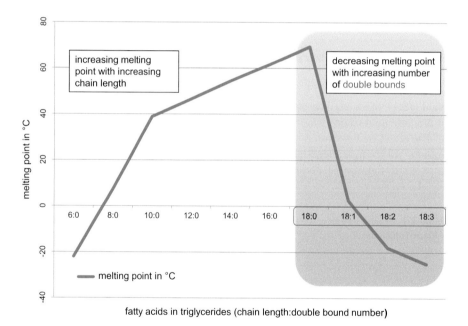

Fig. 4.7 Systematic changes of melting points of triglycerides by carbon chain prolongation and insertion of double bonds

This reduction of melting points due to double bonds in the aliphatic chains has a huge impact as compared to competitive factors. Increasing chain length has an inverse effect, since the longer the aliphatic chain the higher the intermolecular interactions and the higher the melting point. This systematic shift towards higher melting points for longer fatty acid constituents in fats is illustrated in Fig. 4.7. Also the decrease of a melting point by insertion of up to three double points is pointed

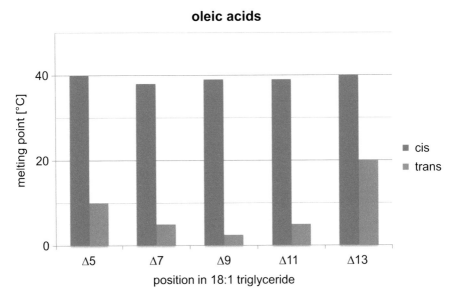

Fig. 4.8 Comparison of melting points for triglycerides with *cis*- or *trans*-unsaturated fatty acids

out for the corresponding C_{18}-acids. The decrease by insertion of the first double
bond (18:1) into a triglyceride corresponds roughly to the increase by prolongation
of the fatty acid chains from C_8 to C_{18}.

The effect by insertion of unsaturation depends also on the double bond configuration, *cis*-configurations have a higher impact as compared to *trans*-configura-
tions due to their lower steric demand (see Fig. 4.6). This relation is pointed out in
Fig. 4.8.

Table 4.6 Typical wax fatty
acids and alcohols

Typical was fatty acids and typical wax alcohols	
$C_{23}H_{47}COOH$	Carnaubic acid
$C_{25}H_{53}COOH$	Cerotic acid
$C_{27}H_{57}COOH$	Montane acid
$C_{15}H_{31}CH_2OH$	Cetyl alcohol
$C_{25}H_{61}CH_2OH$	*n*-hexacosanol
$C_{27}H_{55}CH_2OH$	*n*-octacosanol
$C_{29}H_{59}CH_2OH$	*n*-triacontanol

Fatty acids appear not only in fats but also in further natural products. One group
of biomolecules structurally related to fats are natural waxes. Common and naturally widespread distributed waxes are also esters of long chain acids but linked
with long chain primary mono *n*-alcohols, the so-called fatty alcohols. Fatty acids in
waxes have principally longer carbon chains than fatty acids in fats. Commonly,
chain lengths of the 'waxy' acids are around C_{22} to C_{30}. The corresponding wax
alcohols exhibit similar aliphatic chain lengths (see Table 4.6). In total, waxes can

Table 4.7 Examples of natural and synthetic waxes

Group	Subgroup	Example
Natural waxes	Plant waxes	Candalia wax, carnauba wax, japan wax
	Animal waxes	Bees wax, shellack wax
	Mineral waxes	Ceresin wax, ozokerite wax
	Petrochemical waxes	Paraffin wax
Synthetic waxes		Polyalkylen wax, polyethylene glycol wax

exhibit up to 60 carbon atoms and more. Due to these long aliphatic chains, the waxes exhibit a high lipophilicity and this property is used in nature e.g. for acting as chemical constituents in barriers in organism to control water exchange with the environment. The term waxes does not comprise only the esters of long chain carboxylic acids and long chain *n*-alcohols, but also further waxes of both natural (e.g. ceresin wax) and synthetic origin (e.g. polyethylene glycol waxes) with different chemical properties. Examples are given in Table 4.7. Further on, it has to be noted, that natural waxes normally consist not only of wax esters but represent a complex mixture with further minor ingredients such as *n*-alkanes or *n*-aldehydes.

General Note

Fatty acids appear in nature dominantly as esters with various alcohols. However, the physico-chemical properties of the corresponding natural products are highly determined by the structural features of the fatty acid moieties.

4.3   Phospho- and Glycolipids

Outlook

Phospo- and glycolipids represent a third group of natural products containing fatty acid moieties. Their structural properties are presented.

Beside fats and waxes further natural products exhibit fatty acids as major constituents. Phospho- and glycolipids can be considered as fat derivatives but have quite different functions in organism. A systematic scheme presenting the structural diversity of phospho- and glycolipids is given in Fig. 4.9.

From a structural point of view phospholipids are fat molecules in which one fatty acid moiety has been replaced by phosphoric acid. Since phosphoric acid is a trivalent acid, it can react with further alcohols forming inorganic esters. In natural

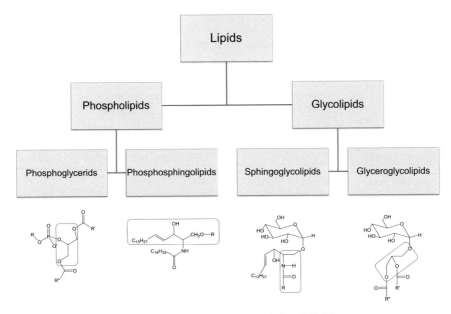

Fig. 4.9 Overview on structural differences of glyco- and phospholipids

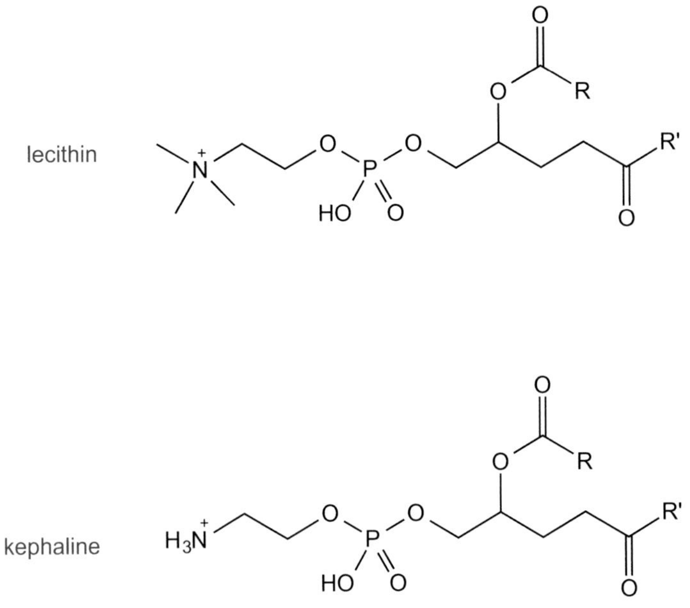

lecithin

kephaline

Fig. 4.10 Chemical structures of some common phospholipids

phospholipids dominantly one further acidic group forms esters with structural diverse alcohols. Main groups belong to aminoethanols or glycerol derivatives. Examples of some major phospholipids are given in Fig. 4.10.

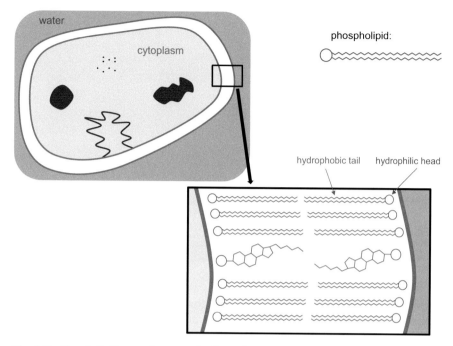

Fig. 4.11 Phospholipids as major parts of cell membranes

Phospholipids are amphiphilic and in aqueous solution they generate a so-called liposome, which is a lipid bilayer. This lipid bilayer points to the biological function of phospholipids, since they are major building blocks of cell membranes with a composition structurally highly related to liposomes (see Fig. 4.11).

Phospholipids also occur as so-called phosphosphingolipids, in which the glycerol moiety is substituted by sphingosine. A similar replacement is observable for glycolipids. Regular glycolipids (glyceroglycolipids) are derivatives of triglycerides by replacing one fatty acid by a sugar moiety linked by an acetale bond (see Fig. 4.9). Analogous, sphingoglycolipids are formed by replacement of glycerol by sphingosine. Glycolipids are also constituents of cell membranes but appear solely at the outer layer.

Excursus: *Phospholipids as indictors for intact microorganism*

Phospholipids underlie a very rapid degradation once released from living cells (e.g. cell death). The detection of phospholipids e.g. in soils is indicative for intact bacteria and, therefore, reflects living microbial activity. This approach allows to use phospholipid analyses for some specific geochemical problems e.g. in areas from which samples for biological analyses are difficult

(continued)

to obtain but for chemical analysis. As an example, Zink et al. (2003) used phospholipid analyses for reflecting active microbial life in marine deep subsurface sediments.

4.4   Fatty Acids in the Geosphere

Outlook

As important parts of more complex biomolecules fatty acids are released to the geosphere and underlie some simple transformation that can be used for characterization of organic matter contribution but also for reflecting thermal maturity.

As described fatty acids exist to a minor extend as free molecules but more important as constituents of various natural products linked via ester bonds. This has some implications for the fate of fatty acids in the geosphere. As a first diagenetic step hydrolysis releases fatty acids from the more complex biomolecules. Hence, the pattern of fatty acids in recent sediments can reflect their original composition in organism. Since some fatty acids are source specific, their occurrence can point to distinct biological input. Such a correlation of fatty acid pattern to source of organic matter is illustrated in Fig. 4.12 by a gas chromatogram as analytical result of fatty acids detection in a river sediment.

This pattern can be interpreted exemplarily as follows:

- The appearance of even C_{12} to C_{18} fatty acids (marked in red) are not specific since they are main constituents of fats and further widespread lipids (phospholipids etc.). This accounts also for the unsaturated acids such as oleic or linolenic acid.
- More specific, the occurrence of *iso-* and *anteiso*-acids (marked in blue) point to contribution of bacterial derived organic matter.
- An elevated concentration of the odd numbered pentadecanoic acid can be related to algae or phytoplankton and characterizes aquatic plant material. This is not too specific, since sediments are part of the aquatic environment.
- The fatty acids with chain length higher than C_{24} derived from cuticular waxes of higher land plants. Therefore, their appearance clearly indicates the contribution of terrestrial organic matter which is an important information regarding the depositional environment.

Since also fatty acids underlie a diagenetic alteration this pattern does not persist in sedimentary systems. The double bonds of unsaturated fatty acids are subject to a

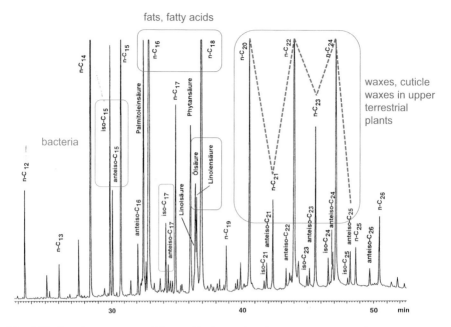

Fig. 4.12 Gas chromatogram of fatty acids detected as their methyl esters in a recent riverine sediment

very rapid alteration. Within the first centimeters of riverine, lake or marine sediments the major proportion of double bonds get lost as depicted by an example in Fig. 4.13.

As a second diagenetic aspect, fatty acids also remain in the sediments in form of their esters. Hence, not only the free fatty acids but also the ester-linked proportion contribute to sedimentary organic matter. From an analytical point of view, the ester-linked fraction can be released by alkaline hydrolysis. The signature of these two fractions can vary significantly (as illustrated in Fig. 4.14) because of the potentially different biological sources.

Excursus: *Traditional soap works*

Alkaline hydrolysis of fats is used by a traditional handicraft since ancient times, the soap work. Soaps have been produced for a long period by applying this reaction on cheap fatty material such as slaughterhouse waste. As a consequence the alkaline hydrolysis of esters is also called *saponification*. Beside glycerol the corresponding fatty acid salts (the soaps) are obtained, which are excellent detergents (see Figure below). Depending on the used base (sodium or potassium hydroxide) two types of soap are differentiated –

(continued)

the curd (sodium salts) and soft soaps (potassium salts, partially with slightly shorter fatty acid chain lengths).

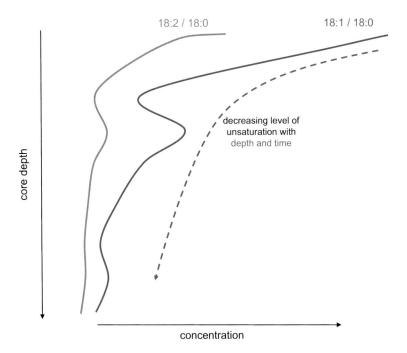

Figure: Saponification of fats

Finally, the most important transformation during diagenesis is the chemical modification of the carboxylic group. As a reaction of minor importance a simple reduction forms corresponding fatty alcohols (see Fig. 4.15). However, the by far more important transformation is the decarboxylation as a form of defunctionalization, leading to *n*-alkanes.

Fig. 4.13 Example for the rapid disappearance of double bonds in sedimentary deposited fatty acids (simplified after Meyers and Ishiwatari 1993)

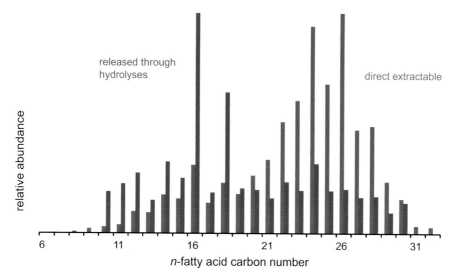

Fig. 4.14 Fatty acid patterns of free and hydrolysable fractions

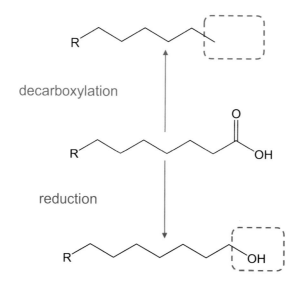

Carbon chain length characteristics		
diagenetic	*n*-alkanes	odd
biogenic	fatty acids	even
biogenic/diagenetic	alcohols	

Fig. 4.15 Principal diagenetic conversions of fatty acids

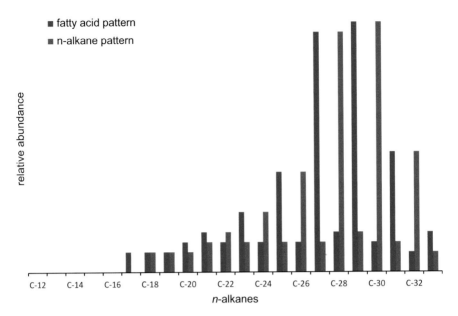

Fig. 4.16 Comparison of fatty acid pattern and diagenetic related *n*-alkane pattern

From an organic-geochemical point of view, two aspects of this diagenetic conversion have a high relevance. On the one hand, *n*-alkanes are very stable and have consequently a high preservation potential in the geosphere. Accordingly, in fossil matter *n*-alkanes can represent an abundant compound group. On the other hand, decarboxylation reduces the number of carbon atoms and transfers the even numbered carboxylic acids to odd numbered *n*-alkanes. This change has to be considered in interpreting fatty acid derived *n*-alkane patterns. An example is given in Fig. 4.16. All interpretations already described above can be transferred to biogenic *n*-alkane pattern but the corresponding chain lengths have to be lowered by one. Therefore, the even-over-odd predominance of biogenic fatty acids is reflected by an odd-over-even predominance for biogenic *n*-alkanes.

Noteworthy, these implications are only valid for biogenic *n*-alkane patterns, that means patterns derived from fatty acid conversion. During diagenesis and more important during catagenesis, this natural signature becomes superimposed by a second fraction, the petrogenic derived alkanes. These aliphatic hydrocarbons are the result of cracking reactions of the macromolecular organic matter, the kerogen, and exhibit no preference neither for odd nor even numbered chain length. An example for a typical petrogenic *n*-alkane pattern as contrasted to a biogenic signature is given in Fig. 4.17.

typical fluvial and oil n-alkanes pattern

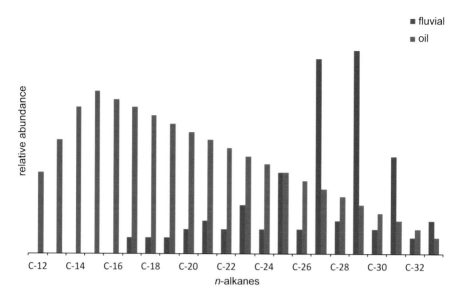

Fig. 4.17 Exemplarily biogenic (fluvial) and petrogenic *n*-alkane pattern in fossil matter

Keeping this superimposition in mind and knowing, that petrogenic signature appear at higher thermal maturity (in the so-called oil window), individual *n*-alkane pattern can be used not only for characterizing the original biogenic contribution in immature organic matter but also for estimating the thermal maturity of fossil material. The continuous superimposition is sketched in Fig. 4.18. With ongoing thermal maturity the original biogenic signature with clear odd-even predominance gets lost and a more unique distribution of chain lengths rises up. At a higher maturity level the biogenic pattern is not visible anymore due to the huge amount of petrogenic formed hydrocarbons.

The changing predominance of chain length is used for calculating the thermal maturity with a very well-known biomarker ratio, the CPI (carbon perforce index). Principally, the relation of odd to even numbered homologues is used, normally for selected chain lengths. The relative abundance is expressed in CPI values that are calculated as relative amount of odd numbered members as compared to even numbered members. CPI values around 1 indicate a uniform distribution reflecting a petrogenic pattern and, consequently, a high thermal maturity. Values significantly higher than 1 point to a more or less significant contribution of the biogenic pattern and a lower maturity.

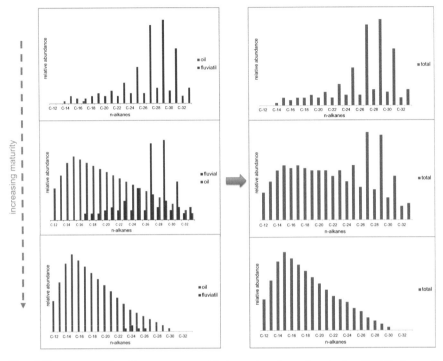

Fig. 4.18 Changes of *n*-alkane patterns with increasing thermal maturity. Contributions of biogenic and thermogenic alkanes are illustrated on the *left side*, the resulting pattern is summarized on the *right side*

4.4.1   Carbon Preference Indices CPIs

$$CPI\ 1 = \ 2 \cdot \frac{nC_{23}+nC_{25}}{nC_{22}+2 \cdot nC_{24}+nC_{26}}$$

$$CPI\ 2 = \ 2 \cdot \frac{nC_{27}}{nC_{26}+nC_{24}}$$

$$CPI\ 3 = \ 2 \cdot \frac{nC_{27}+nC_{29}}{nC_{26}+2 \cdot nC_{28}+nC_{30}}$$

$$CPI\ 4 = \ 2 \cdot \frac{nC_{29}}{nC_{28}+nC_{30}}$$

General Note

It is only a short diagenetic way from fats or waxes to *n*-alkanes. But the geoscientific information derived from *n*-alkane pattern reflecting biogenic input is superimposed by an alternative source of *n*-alkane, the thermal cracking. Knowing this complex situation allows deeper insights into both biogenic origin and thermal maturity of the fossil matter.

4.5   Long Chain *n*-alkenones and Their Organic-Geochemical Relevance

Outlook

A very special application as biomarker is known for long chain *n*-alkenones. These substances are also derived from polyketide biosynthesis and the degree of unsaturation has indicative properties.

A more specific relevance in Organic Geochemistry can be attributed to a further group of natural compounds derived from polyketide biosynthesis. Long chain *n*-alkenones have been identified to be constituents of some marine organism, in particular the diatom *Emiliania huxleyi*, The relevant *n*-alkenones exhibit carbon chain length around C37 to C38 and several double bonds. They appear as methyl and ethyl ketones. The number of double bonds comprises one to four, all with biologically unusual *trans* configuration. Examples of some molecular structures are given in Fig. 4.19.

Interestingly, in the beginning 1980s a systematic shift of degree of unsaturation has been observed for changing water temperatures. The water temperature

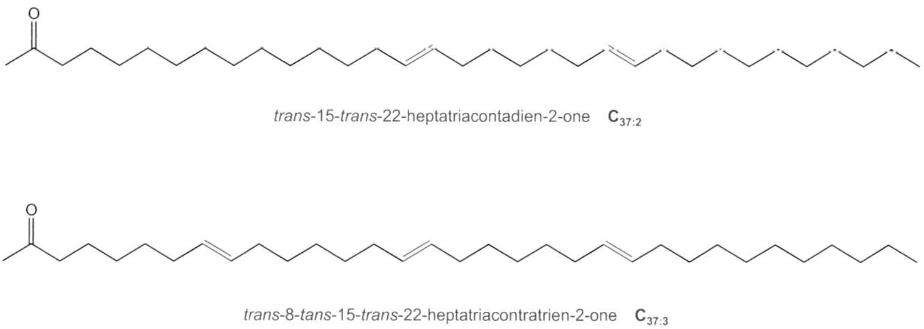

trans-15-*trans*-22-heptatriacontadien-2-one $C_{37:2}$

trans-8-*tans*-15-*trans*-22-heptatriacontratrien-2-one $C_{37:3}$

Fig. 4.19 Molecular structures and shorthand notation

$$U_{37}^{k} = \frac{C_{37:2} - C_{37:4}}{C_{37:2} + C_{37:3} + C_{37:4}}$$

$$U_{37}^{k}$$

$$U_{37}^{k} = 0.040T - 0.11$$

$$U_{37}^{k\prime} = \frac{C_{37:2}}{C_{37:2} + C_{37:3}}$$

$$U_{37}^{k\prime}$$

$$U_{37}^{k\prime} = 0.033T + 0.043$$

Fig. 4.20 Long chain n-alkenone indexes used in Organic Geochemistry (according to Prahl and Wakeman 1987)

influences the biosynthesis and determines the number of double bonds in the long chain n-alkenones. Since this is a systematic shift the relation between degree of unsaturation and water temperature during biosynthesis can be calculated. A corresponding parameter, the so-called U_{37}^{K} index, has been established in Organic Geochemistry because these n-alkenones have been identified in several marine sediment systems. The index is based on relative quantitative distribution of n-heptatriaconten-2-ones. For simplification a shorthand notation according to fatty acids designation is used (se Fig. 4.19). The length of chain as a number is separated by a colon from the number of double bonds and the type of ketone (Me = methyl; Et = ethyl) or the position of the carbonyl group (2- or 3-position), respectively. Generally, the U_{37}^{K} index compares the relative contribution of double unsaturated alkenones relative to triple unsaturated derivatives. Since the 37:4Me isomer was often not detectable, a simplified index was introduced, the so-called $U_{37}^{K\prime}$ index. Both indexes are given in Fig. 4.20.

As a precondition $U_{37}^{K\prime}$ index application in Organic Geochemistry needs a calibration. Such calibrations have been performed in laboratory studies under controlled conditions e.g. with cultures of the diatom *Emiliania huxleyi*. A shift towards a lower quantity of double bonds with increasing temperature is depicted in Fig. 4.21 comparing the relative composition of C37 and C38 alkenones biosynthesized in cultures at 10 °C and 25 °C.

This systematic shift has been also observed in natural systems, e.g. differentiating sediments from temperate or tropical zones. This transferability is illustrated in Fig. 4.22. It is quite obvious, that these variations can be expressed numerically by the $U_{37}^{K\prime}$ index.

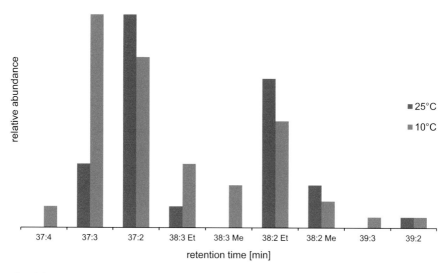

Fig. 4.21 Exemplified patterns of long chain *n*-alkenones in *Emiliana huxleyi* at different temperatures (data adapted from Prahl and Wakeham 1987)

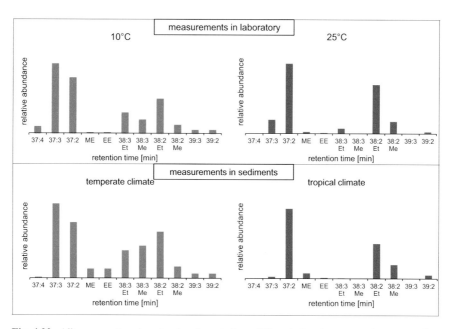

Fig. 4.22 Alkenone pattern in natural sediments from different climate zones and corresponding pattern revealed from laboratory cultures at different temperatures (data adapted from and graphic modified after Prahl et al. 1988)

$U^{K'}_{37}$ indexes calculated from analyses of marine and lake surface sediments and sediment cores have been used intensively as a climate proxy by estimating of water paleotemperatures. The corresponding temperatures represent the water surface temperature, since alkenone biosynthesis by phytoplankton is located in the surface water layers. However, its application is limited in terms of core depth or depositional time since double bonds are sensitive for diagenetic alterations, in particular hydrogenation. Since this process affects the degree of unsaturation, it has huge impact on the calculated indexes. Consequently, the usage of the $U^{K'}_{37}$ index as paleoclimate indicator needs the proof of diagenetically unaltered alkenones.

General Note

Besides GDGTs the alkenones represent a second interesting example of biomarkers directly reflecting paleoenvironmental conditions.

References

Meyers PA, Ishiwatari R (1993) The early diagenesis of organic matter in lacustrine sediments. In: Engel MH, Macko SA (eds) Organic geochemistry – principles and applications. Plenum Press, New York/London, pp 185–209

Prahl FG, Wakeham S (1987) Calibration of unsaturation patterns in long-chain ketone composition forpaleotemperature assessment. Nature 330:367–369

Prahl FG, Muehlhausen LA, Zahnle DL (1988) Further evaluation of long-chain alkenones as indicators of paleoceanographic conditions. Geochim Cosmochim Acta 52:2303–2310

Zink KG, Wilkes H, Disko U, Elvert M, Horsfield B (2003) Intact phospholipids – microbial 'life markers' inamrin deep subsurface sediments. Org Geochem 34:755–769

Further Reading

Rontani JF, Volkman JK, Prahl FG, Wakeham SG (2013) Biotic and abiotic degradation of alkenones and implications for $U^{K'}_{37}$ paleoproxy applications: a review. Org Geochem 59:95–113

Schouten S, Hopmans EC, Sinninghe Damsté JS (2013) The organic geochemistry of glycerol dialkyl glycerol tetraether lipids: a review. Org Geochem 54:19–61

Chapter 5
Pigments

The next important group of compounds used in Organic Geochemistry is not defined by their principal biosynthesis pathway (as for isoprenoids and polyketides) but by their visual properties. Coloring of particular parts of organism is an essential function for visual communication of organism (e.g. to attract or to repel other individuals). This account for both, animals and plants. Color in organism is related to pigments or dyes dominantly belonging to organic substances. The color of organic substances is strongly linked with their molecular structure.

5.1 Relationship of Color and Molecular Structure: The Example of Anthocyanidins and Flavones

Outlook

Although anthocyandins and flavones do not contribute to the field of Organic Geochemistry their molecular structure and its relationship with visible light absorption explains very clearly how organic pigments in plants get their color.

From a physical point of view color is just the process of light absorption or the absorption of photons with energy in the range of visible light. Due to the small size of molecules the absorption has to be described quantum-mechanically. Differences of energy states that are appropriate for light absorption can be associated to the transition of valence electrons between suitable molecule orbitals (transition from ground to an excited state). The energy differences of these orbitals need to fit the energetic spectra of visible light. This is realized by special arrangements in the molecular structure. Most of the valence orbitals

© Springer International Publishing Switzerland 2016
J. Schwarzbauer, B. Jovančićević, *From Biomolecules to Chemofossils*,
Fundamentals in Organic Geochemistry, DOI 10.1007/978-3-319-25075-5_5

differ in their energies by far higher than the energy of visible light
(e.g. differences between σ- and anti-σ-orbitals). Lower energy differences
are realized in π-orbitals as used in double or triple bonds (e.g. the transition
from π- to anti-π-orbitals). However, energy differences of π-orbitals in isolated
double bonds are still to high for absorption in the range of visible light, but
the energy differences are lowered in systems of conjugated double bonds
(alternating arrangement of double and single bonds). The longer the conjugated
system the lower the energy differences and the lower the absorbed energy (or the
longer the wave lengths). Systems with approx. more than 8–10 conjugated
double bonds absorb visible light and are, therefore, colored. Such structural
arrangement, responsible for color in molecules, are named chromophores.
Conjugated double bonds are also an essential component in aromatic systems,
hence also aromatics can contribute to chromophoric systems. Since the color
is correlated with the absorbed energy (according to the energy spectra of
visible light: from higher energy visible in blue to lower energy representing
red), it is also determined by the length of the system of conjugated double bonds.
A secondary element influencing the color are substituents at the chromophores.
Depending on their potential to push or to pull partial electron density, they
shift the color slightly by raising or lowering the energy differences. These effects
are called hypso- or bathochromic shifts, respectively, and depend on the
chemical structure of the substituents.

In plants the highest variety of colors appears in fruits and flowers. For
coloring of these plant parts two organic dye groups are mainly responsible,
the flavones and anthocyanidins. These substances demonstrate nicely how
nature generates the manifold colors in plants. Basic structural elements of
both compound groups are depicted in Figs. 5.1 and 5.2. and a high structural
similarity is obvious.

In both molecular structures aromatic systems, partially with oxygen, build up
the chromophore. Flavones cover yellow to red color, anthocyanidins the color
range from red to blue. As illustrated in Figs. 5.1 and 5.2 the individual colors are
generated by variation of substituents and, in particular, by using the bathochromic
shifts of hydroxy and methoxy groups.

The common names of these pigments point to their occurrence in plants,
e.g. petunidin, malvidin, kaempferol or fisetin. And the tables (Figs. 5.1 and 5.2)
also indicate the widespread appearance of anthocyanidins and flavones in both
flowers and fruits.

Although they occur widespread in plants they exhibit no relevance for
organic-geochemical issues. This is related to their fast degradation as it can be
followed by the decolorisation of fruits and flowers during weathering and decay.
Hence, there is no potential to form chemofossils from anthocyanidines or
flavones.

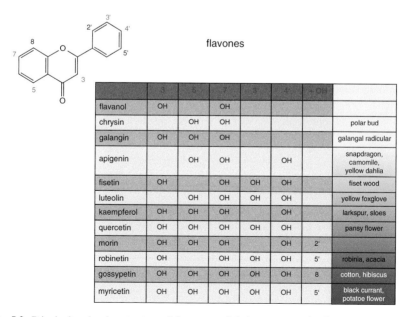

anthocyanidins

	R1	R2	R3	
pelargonidin	H	H	OH	orange dahlia, red currant, blue potaoe flower, Indian cress
cyanidin	OH	H	OH	cherry, cornflower, poppy, plum, red rose
delphinidin	OH	OH	OH	larkspur, purple pansy, kettle hat, lavender,
petunidin	OH	OCH₃	OH	petunia
malvidin	OCH₃	OCH₃	OH	common mallow, blue grape

Fig. 5.1 Principal molecular structure of anthocyanidines and their occurrence in plants

flavones

	3	5	7	3'	4'	+OH	
flavanol	OH		OH				
chrysin		OH	OH				polar bud
galangin	OH	OH	OH				galangal radicular
apigenin		OH	OH		OH		snapdragon, camomile, yellow dahlia
fisetin	OH		OH	OH	OH		fiset wood
luteolin		OH	OH	OH	OH		yellow foxglove
kaempferol	OH	OH	OH		OH		larkspur, sloes
quercetin	OH	OH	OH	OH	OH		pansy flower
morin	OH	OH	OH		OH	2'	
robinetin	OH		OH	OH	OH	5'	robinia, acacia
gossypetin	OH	OH	OH	OH	OH	8	cotton, hibiscus
myricetin	OH	OH	OH	OH	OH	5'	black currant, potatoe flower

Fig. 5.2 Principal molecular structure of flavones and their occurrence in plants

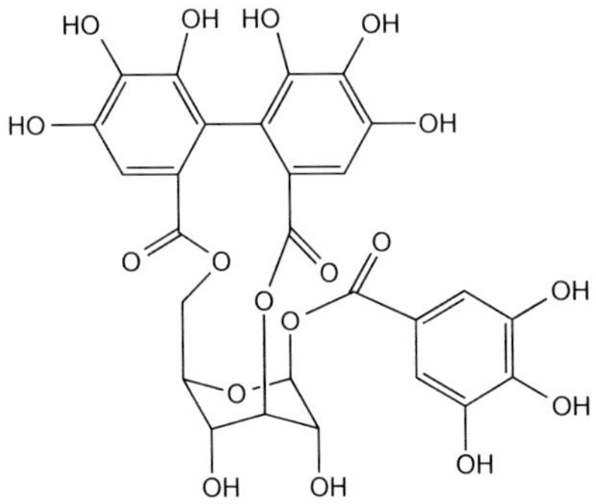

corilagin
- component in pomegranate
- used in traditional Chinese medicine

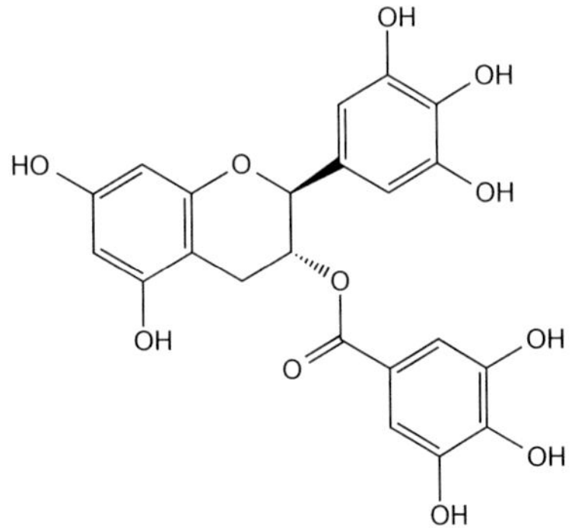

gallocatechin 3-gallate (GCG)
- component of tea
- benefical health effects

Fig. 5.3 Molecular structures of two exemplary tannins

There are certainly many other plant pigments with different chemical structures and biological functions. Tannins for example are known coloring plant constituents. However, their molecular structure is not as systematic as of anthocyanidins and flavones, but can be roughly described as polyphenolic compounds. Some molecular moieties exhibit similarities to the already introduced pigments but the structural variety, in particular with respect to molecular size, is much higher. Some simple examples are given in Fig. 5.3, but many tannins remain structurally unidentified. Although a higher environmental stability is known for tannins, organic-geochemical research on these compound group can be negligible so far.

Higher organic-geochemical importance are evident for two other pigments, carotinoids and chlorophylls, which will be discussed in more detail in the following.

General Note

Color and molecular structure are closely linked. Unsaturation in conjugation is a preriqusite for light absorption and the resulting color can be fine-tuned by substituents. However, color is also a good indicator of diagenetic conversion, since alteration in the chemical structure (either at the chromophore or at its substituents) induces systematic changes in color that can be traced.

5.2   Chlorophylls

Outlook

Chlorophylls are the key molecules in photosynthesis. Their specific molecular structure as well as their appearance in different photosynthesizing organism is introduced.

Beside flowers and fruits also leafs and needles are colored, dominantly in green. This color derives mainly from chlorophylls, which are the key substances in photosynthesis, since they represent the interface for conversion of light energy into chemical energy (see Chap. 1). Chlorophylls are a group of substances structurally conjuncted by a core unit, the so-called porphin or porphyrin. This system consists of four five-membered rings containing one nitrogen atom each. All five-membered rings are connected via methylene bridges to form a superordinated ring system, in which the four nitrogen atoms are aligned to the inner core of the superior ring. Such structure enables a complexation of metal ions by interactions of the free valence electrons as well as the hydrogen atom at the nitrogen with the electron depleted metal ions (Fig. 5.4).

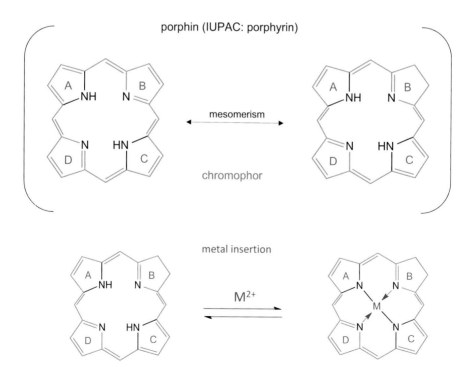

Fig. 5.4 The porphin or porphyrin system and the principal metal insertion reaction

The numerous double bonds in the rings as well as in the bridges are conjugated and, consequently, the porphyrin system represent a chromophor. Its color is influenced not only by the substituents but also by the inserted metal ion. This chromophoric system is used not only in chlorpohylls but also in some other important biomolecules. Two examples for such biomolecules are presented in Fig. 5.5. Haemoglobin as well as cyanocobalamin (vitamin $B_{12,}$ a synthetic precursor of the physiological active vitamin) exhibit a porphyrin unit in which metal ions are enclosed – iron and cobalt, respectively.

Chlorophylls consist also of a central porphyrin unit chelating a magnesia ion and are substituted with some batho- and hypsochromic moieties. A first group of chlorophylls (chlorophylls a, b and d) are connected via an ester bond with an isoprenoidal alcohol, the so-called phytol (see Fig. 5.6). In contrast, chlorophylls c do not contain this phytol moieties (see Fig. 5.7). Bond variation (double or singlebond) at the ethyl substitution attached to the prophyrine system differentiate the 1- und 2-subtype of chlorophyll c.

Commonly, chlorophylls appear in plants as a mixture of a couple of different derivatives. Each chlorophyll exhibit individual absorption ranges and the combination of several pigments allow the absorption of light over a broader range of wavelength. In Fig. 5.8 the total absorption profile is illustrated for a typical higher land plant. Absorption maxima are located in the blue and red wavelength range, so that we see leaves as green, an overlap of the complementary colors.

The occurrence of chlorophylls depends on the type of phototrophs. As listed in Figs. 5.6 and 5.7 chlorophyll a appears in the majority of phytosynthesizing

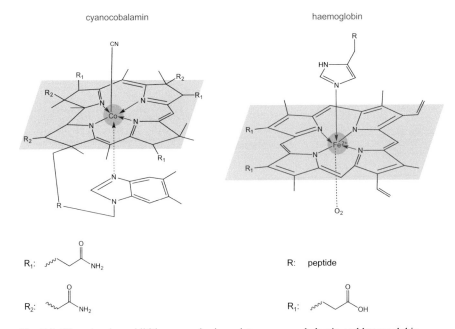

Fig. 5.5 Biomolecules exhibiting a porphyrin moiety – cyanocobalamin and haemoglobin

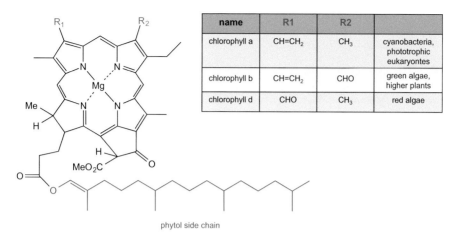

name	R1	R2	
chlorophyll a	CH=CH₂	CH₃	cyanobacteria, phototrophic eukaryontes
chlorophyll b	CH=CH₂	CHO	green algae, higher plants
chlorophyll d	CHO	CH₃	red algae

phytol side chain

Fig. 5.6 Structural diversity of chlorophlylls with phytol side chain

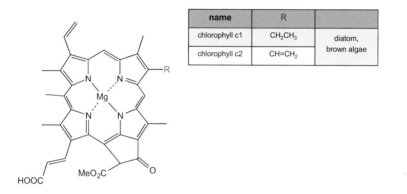

name	R	
chlorophyll c1	CH₂CH₃	diatom, brown algae
chlorophyll c2	CH=CH₂	

Fig. 5.7 Structural diversity of chlorophlylls without phytol side chain

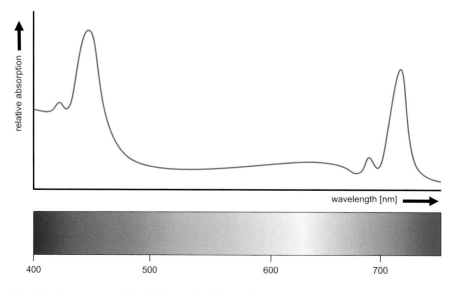

Fig. 5.8 Absorption profile of chlorophylls in higher land plants

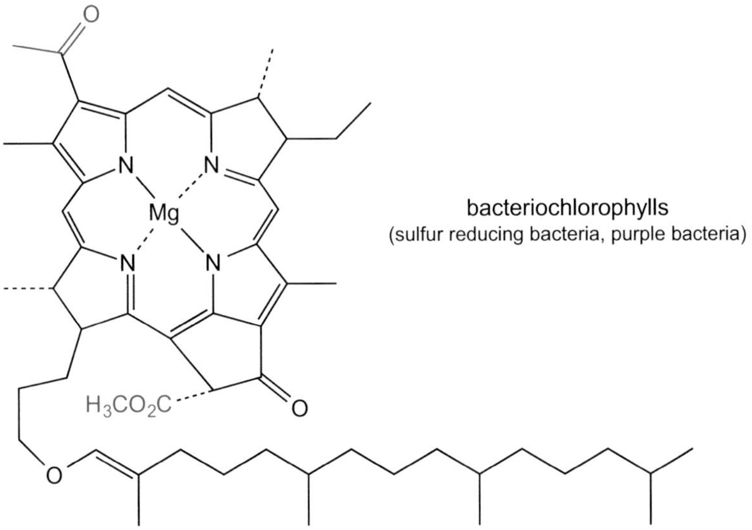

Fig. 5.9 Molecular structure of bacteriochlorophylls

organism, accompanied by the b-type derivative in higher land plants and some algae, whereas the c- and d-type derivatives are constituents in the chloroplasts of more specific types of algae. Lastly, also bacteriochlorophylls exist dominantly appearing in phototrophic bacteria like sulphur reducing or purple bacteria (see Fig. 5.9).

5.3   Fate of Chlorophylls in the Geosphere

Outlook

Both main structural features of chlorophylls, the porphyrin system as well as the phytol side chain are subject to various competitive diagenetic pathways.

Chlorophylls are released to the environment dominantly as a result of segregation after death of phototrophs. They are immediately subjected to conversions that happens in the aquatic environment already in the water phase. However, first alterations can also occur during senescence in the still living organism. All these initial chemical alterations can be followed by the disappearance of the green color as exemplified by Fig. 5.10.

The first diagenetic alterations with organic-geochemical relevance appear already in the water phase and are simply the loss of three significant moieties –

Fig. 5.10 Extracts obtained from aquatic sediments of different ages or first maturity levels, respectively. Note the change of the color from *green* to *brownish* with increasing age (from fresh sediments to aged ones) as the result of vanishing chlorophyll

the phytol side chain, the carboxylic/methylester group at ring C and the central magnesia atom (see Fig. 5.11). As explained before, these units influence the chromophors and, therefore, it becomes obvious why already the initial diagenetic reactions change the color. Initial reaction steps are hydrolysation of the phytol ester and release of magnesia, however their order are not strict and depends partially on the location or initial processes (senescence in organism or segregation after death). Also the third reaction step is not necessarily fixed in order. Therefore, several variations of reaction order exist but all lead to the same intermediate, pyrophaephorbid a.

Noteworthy, the nomenclature of chlorophyll derived diagenesis products is partially systematic but to some extent also confusing. Some prefixes point to the mode of conversion, e.g. 'phaeo' indicates the loss of magnesia and pyro describes the loss of the carboxymethylester moiety at ring C.

After formation of the first relevant intermediate, pyrophaephorbid a, the diagenetic pathway splits again (see Fig. 5.12). These reaction steps are located in the sediments. After a first hydrogenation of the ethylene group at ring A, four reaction steps follow and, as for the first pathway section, the order can vary partially. In summary, the oxo group at ring C becomes reduced, a double bond is inserted at ring D, the carboxylic group of the former phytol ester unit underlies a decarboxylation and, finally, a metal ion enters the free position in the porphyrine ring system. The latter reaction occurs at a later stage of diagenesis/catagenesis, that means at higher maturities. Final product of the total pathway is a metallo deoxophylloerythretioporphyrin (metallo DPEP). Interestingly, the type of inserted metal ions depends on various parameters. During the early diagenesis some environmental parameters like pH and Eh conditions influences metallo porphyrin formation, but also metal availability and competitive reactions (like metal sulfide formation) control the porphyrine metal insertion. Lastly, also maturity determines to some extent the metal insertion. Generally, the following bivalent cations have been identified as major constituents of metallo porphyrines: VO^{2+} (vanadyl oxide), Ni^{2+}, Cu^{2+}, Fe^{2+} and Zn^{2+}. The process of incorporation is not fully understood. It

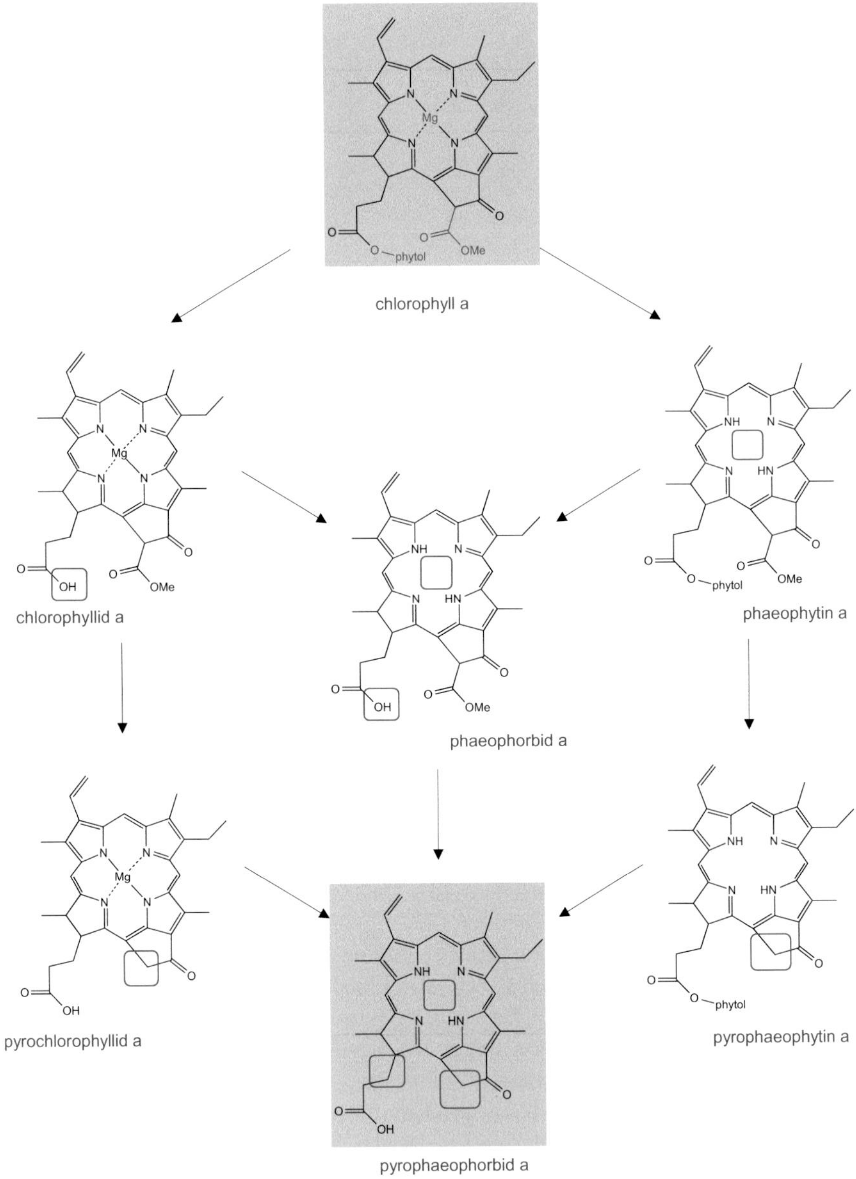

chlorophyll a

chlorophyllid a

phaeophytin a

phaeophorbid a

pyrochlorophyllid a

pyrophaeophytin a

pyrophaeophorbid a

Fig. 5.11 Simplified first diagenetic reaction steps of chlorophyll a, focusing on the prophyrine moiety. This modification take place dominantly in the water column (simplified and modified after Killops and Killops (2005) and references cited therein, Junium et al. 2015)

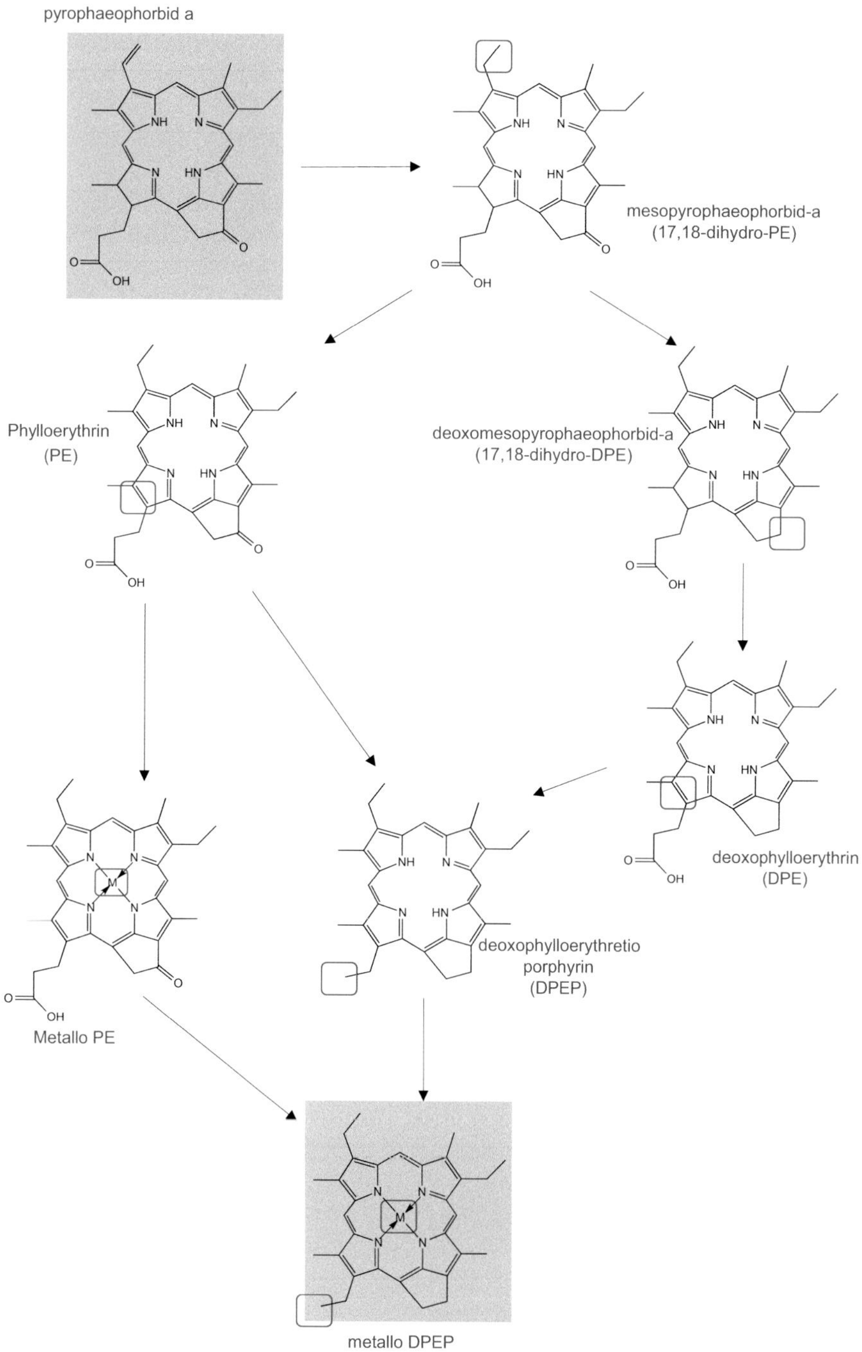

Fig. 5.12 Second part of the simplified diagenetic and catagenetic pathway of the chlorophyll derived prophyrine moiety (simplified and modified after Killops and Killops (2005) and references cited therein, Junium et al. 2015)

seems that there is a kind of order of incorporation, but not too systematically. A rough incorporation order might be: $Cu^{2+} < Ni^{2+} < VO^{2+}$. For nickel a relatively clear correlation with depth or maturity is obvious. Interestingly, Ni and V porphyrines can be differentiated by their accessibility. This is illustrated by an analytical approach: Ni porphyrines are extractable whereas V porphyrines are a non-extractable species.

Also for the second part of the pathway a semi-systematic nomenclature for the intermediates is used. Prefixes are used to characterize the double bond hydrogenation (meso), ketone reduction (deoxo) and metal ion insertion (metallo).

To have a more general overview on this complex pathway, a strongly simplified sketch is given in Fig. 5.13, in which the connection of the more water related and more sediment located modifications as well as the role of pyrophaeophorbid a as connecting derivative are figured out.

The diagenetic pathway of chlorophyll is an excellent example to point to one aspect that is often disregarded. The fate of organic molecules in the aquatic environment and the corresponding sediment systems is generally not represented by a singular diagenetic pathway. Organic-geochemical research focuses dominantly on those molecular residues that survived over geological time (as pointed out in Chap. 2). However, the main proportion of biomolecules becomes mineralized or degraded to a huge extent. Further on, individual molecules can underlie different and competitive reaction courses. With respect to chlorophylls such alternative routes are known already at the beginning of diagenesis. The prophyrine

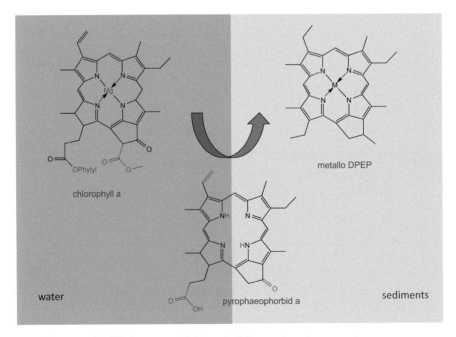

Fig. 5.13 Very simplified summary of chlorophyll diagenesis and catagenesis

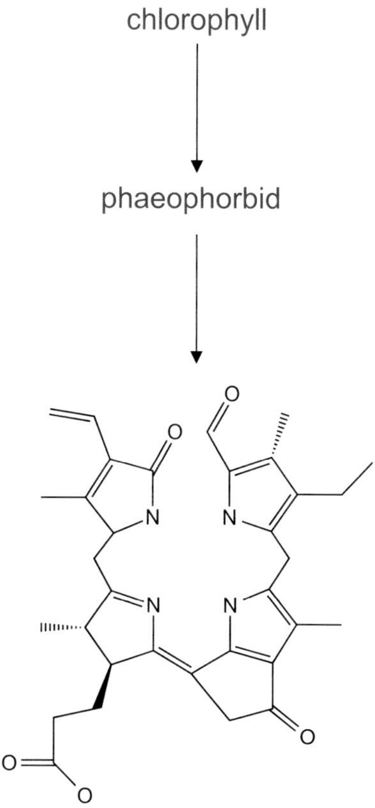

Fig. 5.14 Porphyrine
system cleavage as
alternative degradation
pathway of chlorophylls

system is not as stable as it might seems so far. A major proportion of released
chlorophyll is subject to ring system cleavages. Since this reaction enhances further
degradation steps and, additionally, alters the molecular structure to a high extend
(preventing a preservation of structural specifity), this reaction pathway is
uninteresting from an organic-geochemical point of view (Fig. 5.14).

Excursus: *Phorphyrine ring system cleavage in nature*

A ring cleavage degradation is also known for porphyrine derivatives in
humans. Haem as part of haemoglobin, an essential blood constituent already
depicted in Fig. 5.5, is decomposed in our bodies by ring cleavage in the liver
forming bilirubin, which is transformed by intestinal bacteria to urobilinogen
and, following, is excreted. Since urobilinogen is unstable against oxidation,
it is transformed after excretion to urobilin. This substance acts as anthropogenic tracer in environmental studies to follow fecal emissions in the aquatic
environment (Takada and Eganhouse 1998).

(continued)

Organic-geochemically more relevant are alternative routes including formation and cleavage of ring structures. It is known, that phaeophytin a can react (preferentially in carbonates under anoxic conditions) at ring D via a type of intramolecular aldol condensation to form a seven-membered ring. This ring survives the further diagenetic reactions and appears as specific structural moiety in the final products the bicycloalkanoporphyrines BiCAP (with or without metal ion inserted) (Fig. 5.15).

A second alternative diagenesis route is characterized by a ring cleavage (see Fig. 5.16). At the five-membered ring attached to ring C the neighbored oxo and carboxylic moieties destabilize the bridging bond. Therefore, an intermolecular rearrangement can form a carboxylic anhydride group which can be easily cleaved by hydrolysis. The resulting dicarboxylic acid undergoes a double decarboxylation resulting finally in a complete removal of the former five-membered ring. These final products obtained after subsequent defunctionalisation are named etioporphyrins or metallo etioporphyrins, if metal ions are inserted.

Noteworthy, the appearance of etioporphyrins in fossil matter is not necessarily the result of chlorophyll diagenesis as described in Fig. 5.16. Conversion of

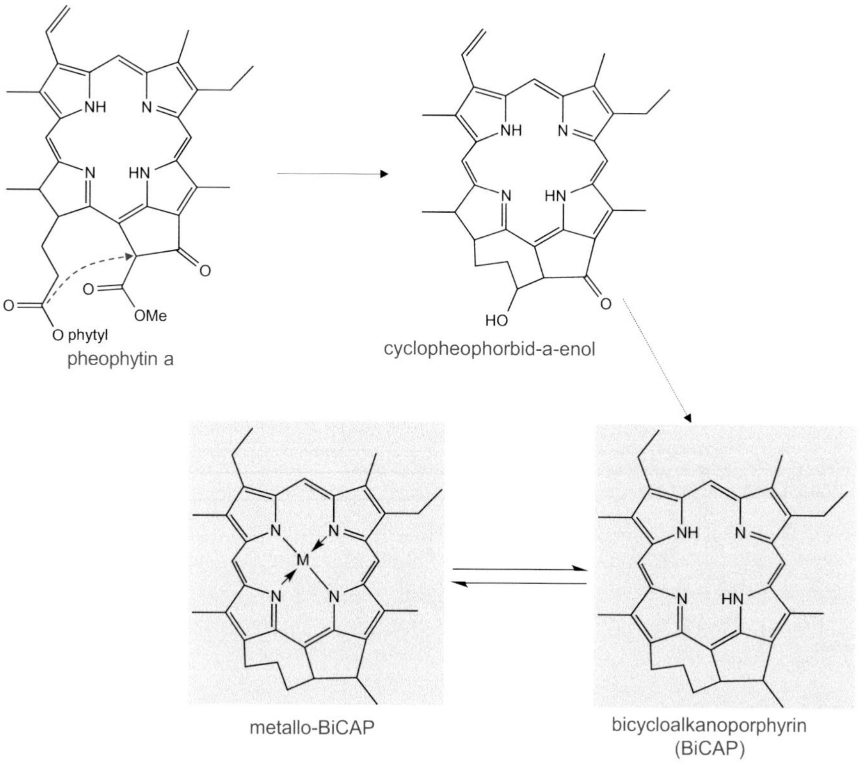

Fig. 5.15 Cyclisation at ring D as alternative route within the chlorophyll diagenesis pathway (according to Junium et al. 2015)

alternative biomolecules with similar structure such as haem can result in etiophorphyrin products and also alternative diagenetic pathways like the ring cleavage of metallo DPEP as a final transformation may produce desethyl etioporphyrins (see Fig. 5.17).

> **General Note**
>
> Chlorophyll derived phorphyrin moieties undergo a complex set of diagenetic pathways which appear competitively in nature. However, the characteristic porphyrin system remains often as indicative structural element in the chemofossils.

A second important aspect of chlorophyll diagenesis has been neglected so far. The phytol side chain (its isoprenoidal origin has been already introduced in Chap. 3) has its separated and unique fate in the geosphere. Interestingly, after release from the chlorophyll core the reaction pathway differs according to the

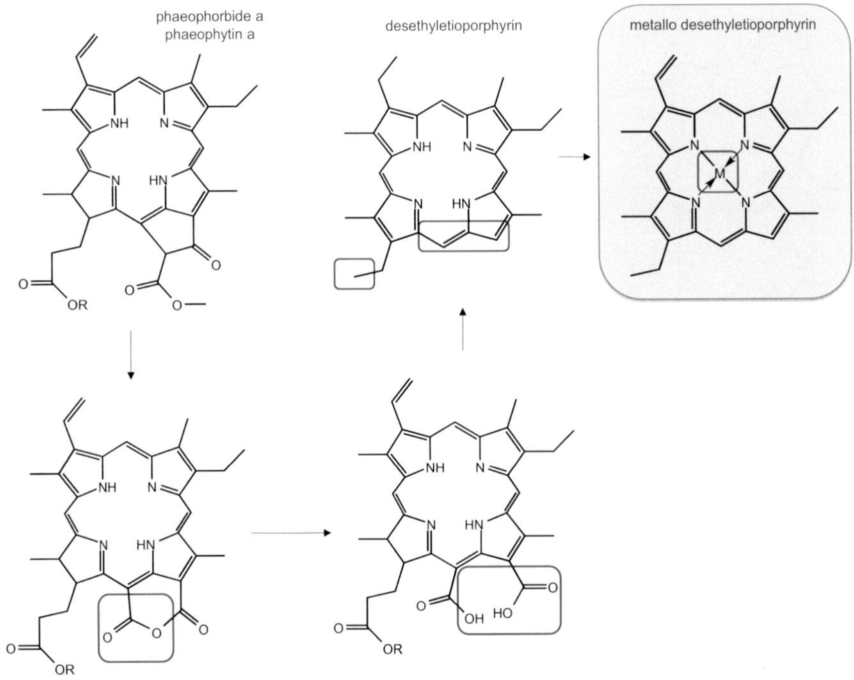

Fig. 5.16 Ring cleavage near ring C as alternative route within the chlorophyll diagenesis pathway (simplified and modified after Killops and Killops 2005)

depositional conditions. However, the resulting pathways remain simple as depicted in Fig. 5.18.

The principal reaction type is defunctionalization. Under strict anoxic conditions the hydroxyl group and the double bond of phytol get reduced to form (via dehydrophytol) a simple aliphatic hydrocarbon, the so-called phytane. The chiral atoms (as result of the characteristic isoprenoid methylsubstitution pattern) remain with their biogenic 'imprint' at first before an epimerization can be observed under catagenetic conditions. Under more oxic conditions phytol can become oxidized resulting in the corresponding phytenic acid, but following decarboxylation forms also an aliphatic hydrocarbon, the so-called pristane. This diagenetic product exhibits one carbon atom less as compared to phytol and phytane. Also the stereoisomers (note: only two chiral atoms remain) become epimerized under higher pressure and temperature. Therefore, the relative composition of pristine and phytane diastereomers are used as maturity marker. Though, the main application of pristane and phytane is related to their formation under different depositional conditions. It has been demonstrated that salinity and oxygen availability during deposition determine the extent of phytane and pristane formation. Hence, in turn the relation of these two diagnetic products is indicative for the depositional

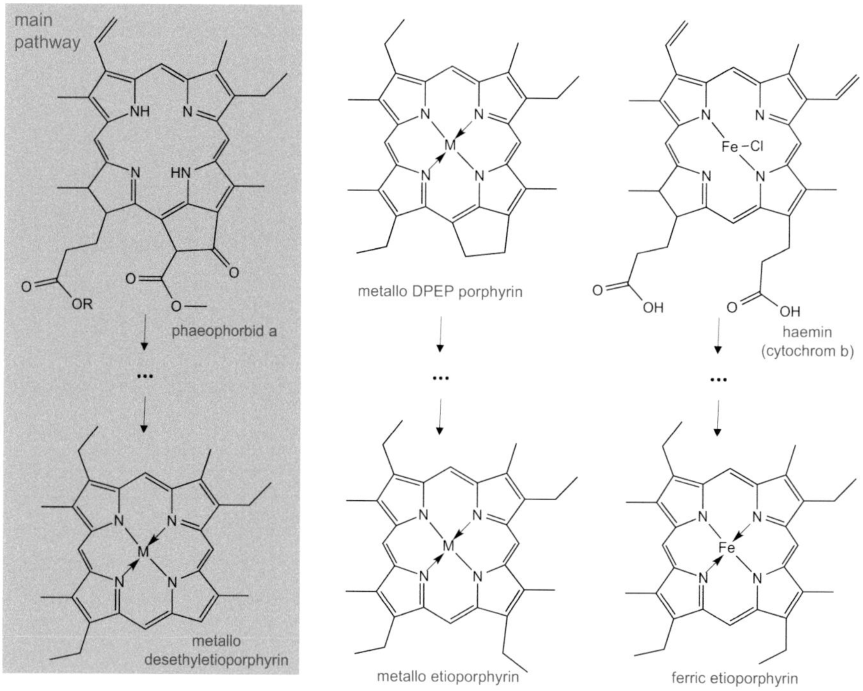

Fig. 5.17 Alternatives for desethyletioprophyrine formation (simplified after Killops and Killops)

environment. This approach is commonly used after normalization of the individual isoprenoic hydrocarbons to their gas chromatographically co-eluting n-alkanes (namely n-heptadecane, n-C_{17}, and n-octadecane n-C_{18}). Usually a two dimensional correlation is used for graphic interpretation of these biomarker ratio as illustrated in Fig. 5.19. Beside information about the depositional environment also the type of organic matter can be figured out. However, the efficiency of this biomarker ratio is partially lowered by alternative sources in particular for the appearance of phytane. Under certain conditions phytane may derive also from other biomolecules such as tocopherol (or other chromanes with isoprenoic side chains) or from cracking reactions of isoprenoic kerogen moieties.

As last information a shift towards higher relative proportion of pristane and phytane as compared to the n-alkanes points to biodegradation processes. This is based on the observation that non-branched alkanes are subject to more intensive microbial degradation as compared to branched aliphatic hydrocarbons, which induces a systematic relative decrease of n-alkanes with ongoing biodegradation. On the contrary, an opposite shift towards lower pristane and phytane rations points to a relative increase of the n-alkanes, which can be explained only by thermogenic generation (cracking). Consequently, this shift correlates with maturity.

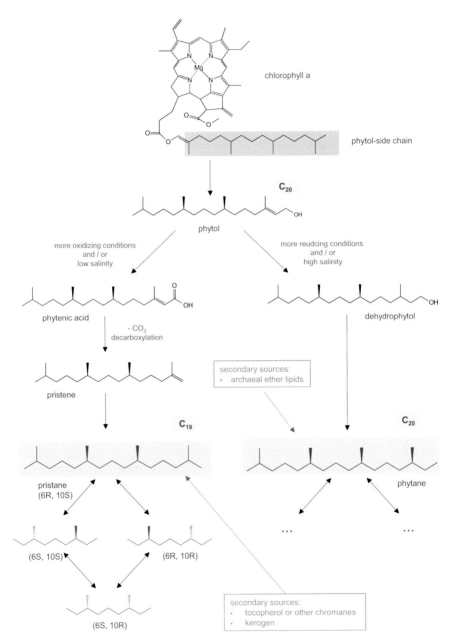

Fig. 5.18 Main diagenetic pathways of phytol (modified and simplified after Didyk et al. 1978; Rontani and Volkman 2003; Killops and Killops 2005)

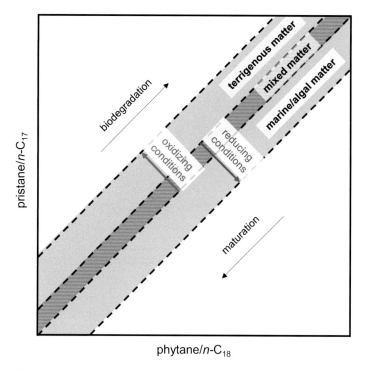

phytane/n-C$_{18}$

Fig. 5.19 Interpretation scheme for the comparison of phytane to pristane abundance as indicator for depositional conditions and environment as well as biodegradation (according to Hunt 1996)

General Note

Pristane and phytane as main diagenetic products of phytol are used as a manifold biomarker system. Thermal maturity, depositional conditions and biodegradation are reflected sensitively by these chemofossils.

Case Examples: Pri/phy ratio as redox parameter

The relative proportion of pristane and phytane expressed as pri/phy ratio has been used frequently to characterize changes in oxygen availability during sedimentation conditions. Two examples are figured out below reflecting a transition from aerobic to strongly anaerobic conditions in Lower Toarcian shales (Moldowan et al. 1986) and the Permian-Triassic superanoxic event

(continued)

(Grice et al. 2005). Noteworthy, this parameter has been used complementary with many other indicators (e.g. Ni/V ratio, sulphur isotope data, monoaromatic sterane ratios etc.) since one single parameter is not able to identify unambiguously such transitions of paleoenvironmental conditions.

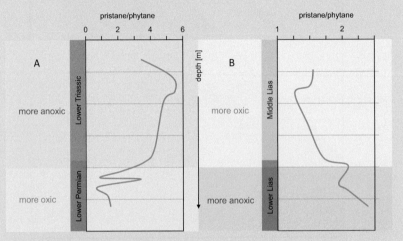

Figure: Course of pri/phy ratios reflecting two different transitions either from oxic to anoxic conditions (A) or vice versa (B), data adopted from Grice et al. (2005) and Moldowan et al. (1986), respectively

5.4 Carotenoids

Outlook

Carotenoids are tetraterpenes with some variations in their molecular structure. This diversity as well as the main diagenetic pathways are discussed.

A second group of organic pigments exhibiting geochemical relevance are the carotenoids. These biomolecules appear like chlorophyll widespread in plants from both the marine and terrestrial environment. They absorb light in the violet to green range resulting in a complementary color of orange to red. With light absorption carotenoids support chlorophyll in photosynthesis in various ways (absorption, protection etc.). In higher animals carotenoids act in diverse functions also related dominantly to their ability of light absorption.

In Chap. 3 carotenoids have already been introduced as representatives of tetraterpenes (8 isopren units, 40 carbon atoms). The chromophor consists of numerous conjugated double bonds along the carbon chain. Carotenoids appear in organism with several derivatives. Variations are located at the end groups

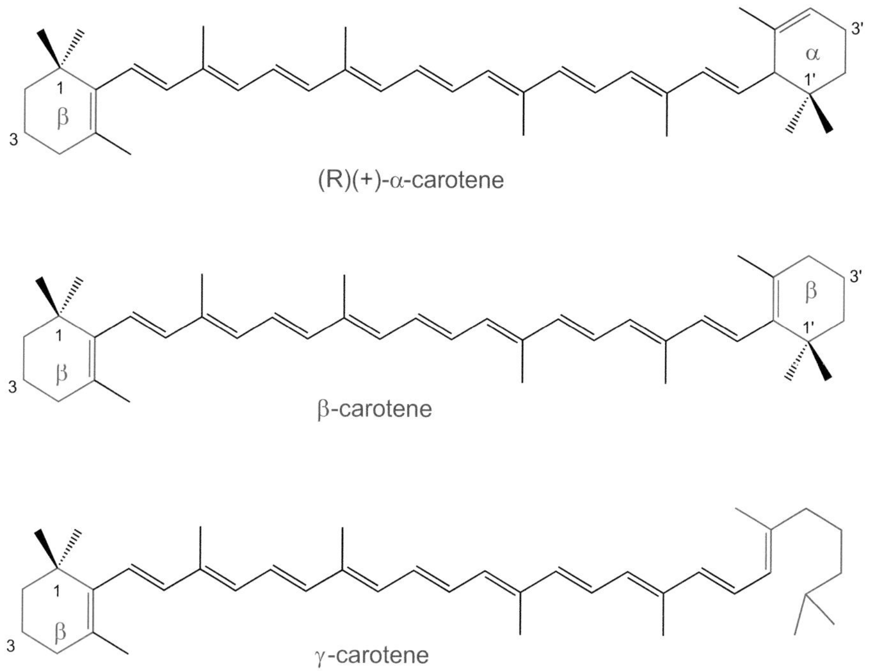

(R)(+)-α-carotene

β-carotene

γ-carotene

Fig. 5.20   Molecular structures of some carotenes

comprising cycles or acyclic moieties as well as different locations of double bonds (see Fig. 5.20). Principally, two different ring systems, labelled with α or β, and an open structure (γ) differentiate α-, β- and γ-carotenes.

Further on, functionalized carotenoids (dominantly with oxygen containing groups, see Fig. 5.21) are called xanthophylls, whereas pure hydrocarbons are named carotenes.

Also the diagenesis of carotenoids consist of various pathways. A very early conversion (just in the water phase and surface sediments) is known resulting in the formation of loliolide and related derivatives (see Fig. 5.22). As a first reaction step the epoxidation of the ring double bond leads to two different stereoisomers, depending on the direction of the electrophilic attack. Subsequent cyclisation involving the first double bond of the acyclic chain and forming an oxygen containing five-membered ring as well as the loss of the side chain results in the formation of loliolide and isololiolide. Both compounds just differ in the stereochemical configuration at positions 3 and 5 as the result of the first reaction step. These degradation products (as well as derivatives with other substitution patters at the ring system, e.g. no hydroxy group in 3-position, resulting finally in dihydroacinidiolide) can be often detected in recent sediments representing the contribution of plant derived organic matter.

A higher geochemical importance exhibit other diagenetic products, the saturated hydrocarbons generated by defunctionalisation (in the case of xanthophylls)

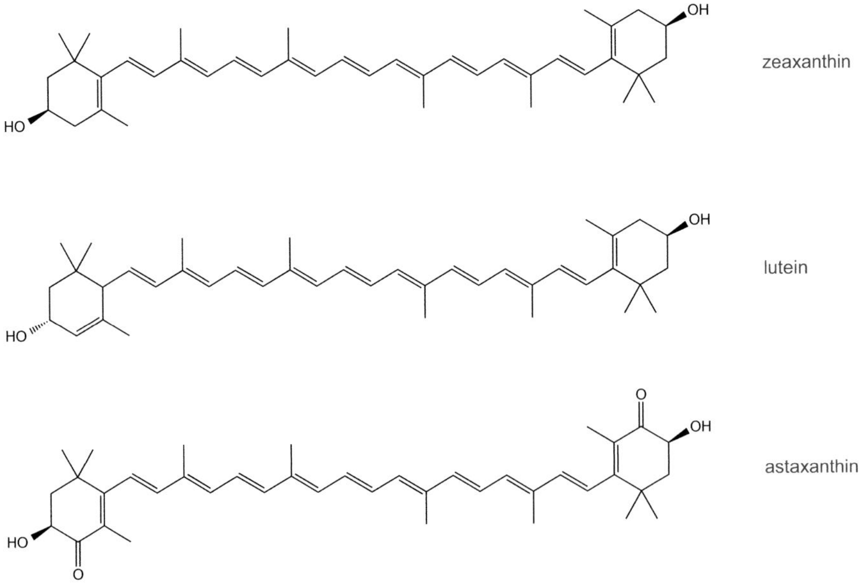

Fig. 5.21 Molecular structure of selected xanthophylls

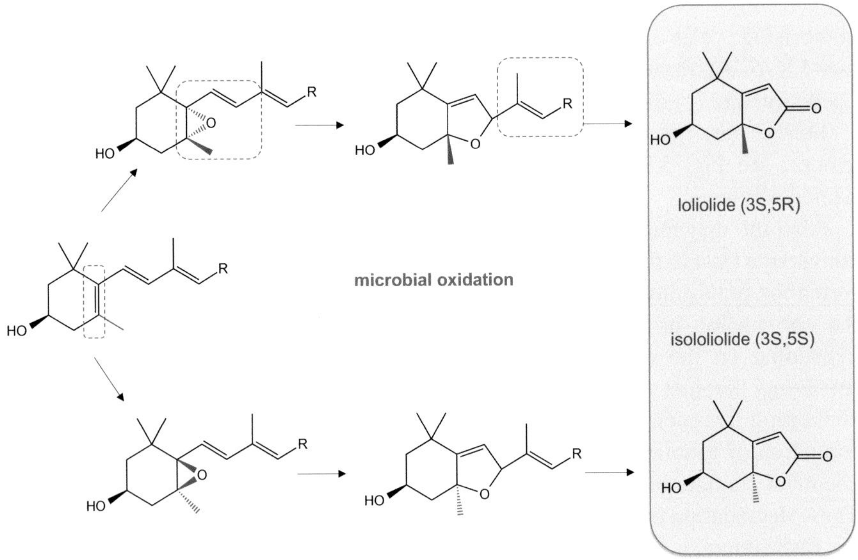

Fig. 5.22 Early diagenesis of carotenoids forming bicyclic products (adapted and simplified after Killops and Killops 2005)

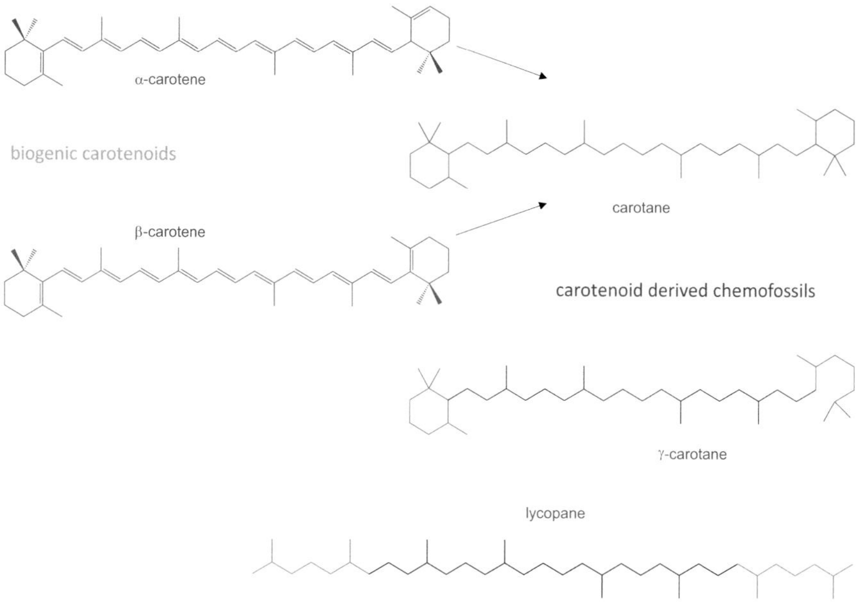

Fig. 5.23 Carotanes and their structural information of former carotenes

and hydrogenation of all double bonds. These carotanes are stabilized and, concurrently, preserve some specific structural units pointing to carotenoid origin. This applies especially for the structure at the chain ends, where former cyclisation patterns can be recognized as pointed out in Fig. 5.23. However, the preservation is certainly limited, e.g. former α- and β-carotenes cannot be distinguished by their diagenetically formed carotanes, since their structural difference (location of the ring double bond) vanished.

Besides the generation of aliphatic hydrocarbons some interesting diagenetic reactions are known for carotenoids forming aromatic moieties. Firstly, an aromatic structure is already pre-build at the cyclic end groups of α- and β-type carotenoids. However, the geminal methyl substitution prevents an easy aromatization due to the quaternary central carbon atom. However, the 'energetic benefit' of aromatization leads to a rearrangement with a migration of one geminal methyl group to a neighbor position (see Fig. 5.24, change from 2,2,6- to 2,3,6-trimethylsubstitution pattern). This rearrangement opens the way to form aromatic rings at the end positions of carotenes. Products are isorenieratane and related derivatives.

Secondly, also along the acyclic mid chain with its high number of conjugated double bonds intramolecular cyclisation reactions can occur leading preferentially to cyclohexadiene moieties as illustrated in Fig. 5.25. With a last dehydrogenation step aromatic rings can be formed.

These cyclisation reactions are not restricted to defined positions at the chain and are not limited to monoaromatic rings. As a consequence numerous individual

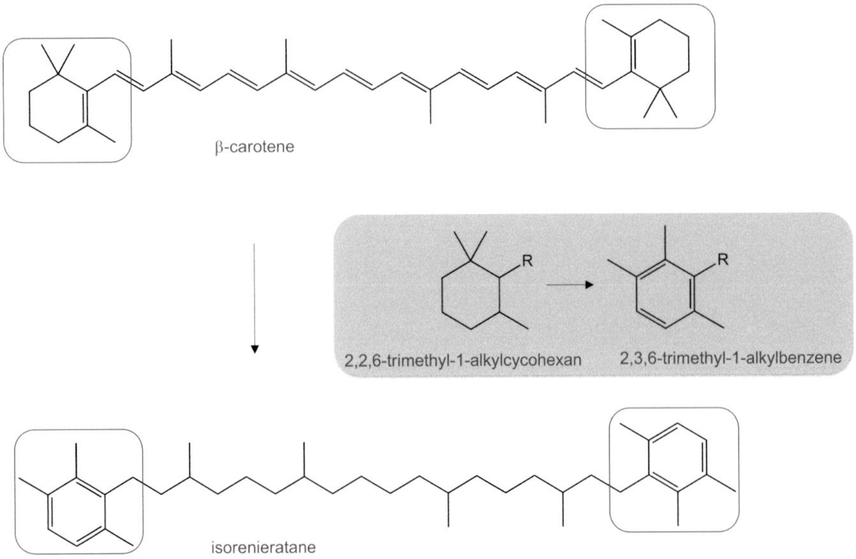

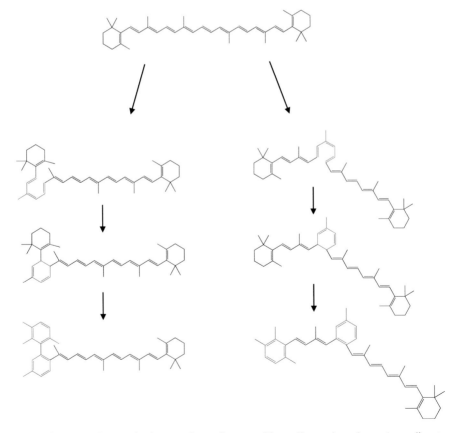

β-carotene

2,2,6-trimethyl-1-alkylcycohexan 2,3,6-trimethyl-1-alkylbenzene

isorenieratane

Fig. 5.24 Formation of aromatic rings at the end positions of carotenes

Fig. 5.25 Internal aromatization reactions of carotenoids as diagenetic pathway (according to Koopmans et al. 1997)

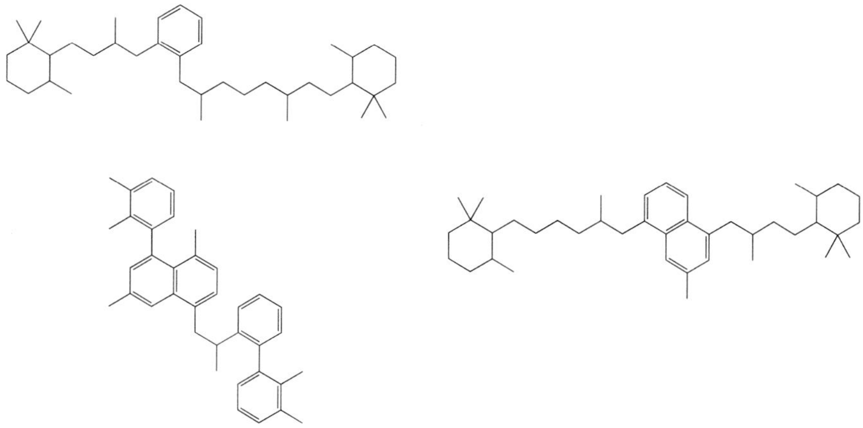

Fig. 5.26 Selected carotene derived compounds with internal aromatic rings (according to Koopmans et al. 1997)

carotene derived compounds with internal aromatic moieties have been identified as exemplified in Fig. 5.26.

> **General Note**
>
> For carotenoids the aromatization either at the already existing end ring systems or at internal parts of the chromophoric system is a main path for stabilization. The isoprenoidal basic structure allows a classification of the corresponding chemofossils as carotenoid derived biomarker.

References

Didyk BM, Simoneit BRT, Brassel SC, Eglinton G (1978) Organic geochemical indicators of palaeoenvironmental conditions of sedimentation. Nature 272:216–221

Grice K, Cao C, Love GD, Böttcher ME, Twitchett RJ, Grosjean E, Summons R, Turgeon SC, Dunning W, Jin Y (2005) Photic zone euxinia during the Permian-Triassic superanoxic event. Science 307:706–709

Hunt JM (1996) Petroleum geochemistry and geology, 2nd edn. Freeman, New York

Junium CK, Freeman KH, Arthur MA (2015) Controls on the stratigraphic distribution and nitrogen isotopic composition of zinc, vanadyl and free base prophyrins through Oceanic Anoxic Event 2 at Demerara Rise. Org Geochem 80:60–71

Killops S, Killops V (2005) Introduction to organic geochemistry, 2nd edn. Blackwell Publishing, Oxford

Koopmans MP, de Leeuw JW, Sinninghe Damste JS (1997) Novel cyclized and aromatized diagenetic products of β-carotene in the Green River Shale. Org Geochem 26:451–466

Moldowan JM, Sundararaman P, Schoell M (1986) Sensitivity of biomarker properties to depositional environment an/or source input in the Lower Toarcian of SW-Germany. Org Geochem 10:915–926

Rontani JF, Volkman JK (2003) Phytol degradation products as biogeochemical tracers in aquatic environment. Org Geochem 34:1–35

Takada H, Eganhouse RP (1998) Molecular markers of anthropogenic waste. In: Meyer RA (ed) Encyclopedia of environmental analysis and remediation. Wiley, New York, pp 2883–2940

Chapter 6
Macromolecules

6.1 Principals

Outlook

Basic aspects of the structural properties of synthetic and natural polymers are summarized. Focus lies on the type of linkage and the resulting polymer properties.

Low molecular weight compounds with lipophilic character and stable enough to survive in the geosphere over a long time can act as biomarker as pointed out in Chap. 2. However, the dominant proportion of biogenic organic matter belongs to the group of macromolecular or high molecular weight substances. The systematic description of these substances differs as compared to the chemical characterization of small molecules. Generally, subunits and their type of repetitive linkage are the clue to understand the structure and composition and, consequently, the chemical behavior of macromolecules. Generally, the boundary between low and high molecular weight substances is not keen but more a fluent transition. Roughly, compounds with up to 1000 Da belong to the low molecular weight fraction, whereas high molecular weight substances can exhibit molecular masses up to several million Dalton (see Fig. 6.1).

The best way to introduce this approach is to begin with the synthetic macromolecules or polymers. The technical synthesis of polymers started in the end of the twentieth century. Firstly, natural polymers have been chemically modified, e.g. celluloid or viscose (see Table 6.1).

Following, full synthetic polymers have been introduced in the beginning of the twentieth century starting with the synthesis and production of bakelite (see Table 6.2). In modern times synthetic polymers play an important role in everyday

© Springer International Publishing Switzerland 2016
J. Schwarzbauer, B. Jovančićević, *From Biomolecules to Chemofossils*,
Fundamentals in Organic Geochemistry, DOI 10.1007/978-3-319-25075-5_6

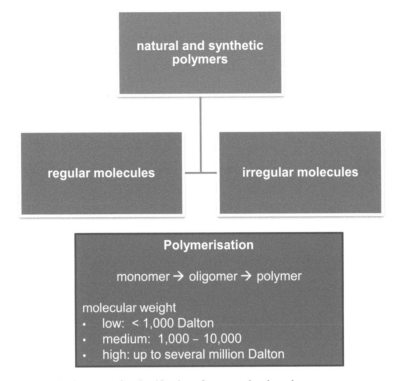

Fig. 6.1 Some basic aspects for classification of macromolecular substances

Table 6.1 First semi-synthetic polymers and their technical applications

Semi-synthetic plastics		
Year	Substance	Application
1896	Celluloid (cellulose nitrate + camphor)	Ivory substitute (billiard ball), film material, TT-balls, combs
1882	Viscose (cellulose + NaOH + CS$_2$)	Textile fibers, cellophane plastics
1897	Galalithe kunsthorn (casein + formaldehyde)	Studs, knife handles

life and common derivatives are well known to everyone such as polyethylene, polystyrene or nylon.

The basic composition can be described best by the principal synthesis route. Low molecular weight units, the so-called monomers (*mono* is the ancient Greek word for 'single'), are connected repetitively via two principal modes of linkage. The connection of a few monomers forms so-called oligomers (*oligo* is the ancient Greek word for 'few'), but the multiple conjunctions of monomers composes polymers (*poly* is the ancient Greek word for 'many' or 'multiple'). The principal modes of linkage are polyaddition and polycondensation. Polyaddition is the repetitive addition of monomers dominantly by cleavage of double bonds and

Table 6.2 Synthetic polymers and their technical application

Synthetic plastics		
Year	Substance	Application
1907	Bakelite	Receiver cabinet, insulators
1933	Plexiglass	Synthetic glass
1933	Low density polyethylene	Diverse
1935	Nylon	Textile fibers
1936	Polystyrene	Diverse
1937	Polyvinyl chloride	Flooring
1938	Perlon	Textile fibers
1953	High density polyethylene	Diverse
1953	Polycarbonates	CDs
1957b	Polypropylene	Diverse

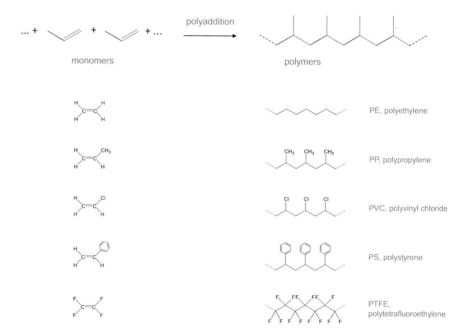

Fig. 6.2 Scheme of polyaddition and some examples of synthetic polymers formed by polyaddition

subsequent forming of new intermolecular bonds. The principal reaction as well as some examples of synthetic polymers are given in Fig. 6.2.

A second type of polymer formation is realized by multiple condensation reactions. For this purpose appropriate functional groups are needed, with which a connection under release of a small molecule as side product (normally water) can be created. Common condensation reactions used for polymer synthesis are the formation of esters and amides connecting carboxylic groups with hydroxyl or

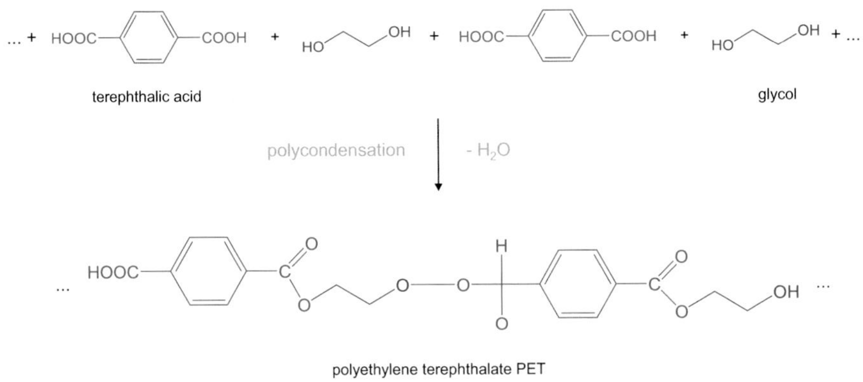

Fig. 6.3 Formation of poly ethylene terephthalate (*PET*) by polycondensation of terephthalic acid and glycol. The polymer is built up by polyester linkages

amine groups. The polymer products are named polyesters or polyamides. Noteworthy, such condensation reactions are commonly reversible. An example for a polycondensation reaction is given in Fig. 6.3.

Macromolecular substances can be further classified by their regularity of monomer addition. If a very strict periodical connection of the monomers is realized, a very uniform polymer is formed, a so called regular polymer (see Fig. 6.1). However, in particular as a result of polyaddition the formation of polymer linkages can be more erratically and more irregular polymers can be formed.

From a geochemical point of view, synthetic polymers exhibit no relevance. But biogenic macromolecules, or biopolymers, represent the major proportion of organic matter in the biosphere. And geogenic macromolecules dominate the organic fraction in the geosphere as well. In more detail, organisms are principally composed dominantly by biopolymers like proteins, polysaccharides or lignin (see Fig. 6.4). This accounts for simpler organism such as bacteria as well as for more complex organism like higher land plants.

The fraction of lipids have been discussed intensively in the last chapters due to their importance as biomarkers. However, these compounds contribute only to a low extend to the biogenic organic matter. Therefore, due to its high abundance the macromolecular organic matter seems to be very interesting for Organic Geochemistry. Similar to the criteria applied to low molecular weight substances the potential to exhibit organic geochemical relevance is not solely restricted to an abundant occurrence in the biosphere but also to the preservation of the compounds in the geosphere over long time periods. And in particular for individual biopolymers the preservation potential differs enormously. This is summarized in Table 6.3, where a low preservation on a geological time scale is evident for the better known biopolymers, like proteins, polysaccharides as well as DNA. But a geochemically sufficient preservation is obvious for some less known biopolymers like suberin, algaenan or lignin.

| | | | proteins | polysaccharides | | lignin | lipids |
				cellulose	others		
higher land plants	leaves Species	leaves					
		wood					
	coniferous species	neadles					
		wood					
lower land plants	mosses						
	lichen						
higher water plants	potamogaton						
	phragmites						
phytoplankton	blue-green algae						
	diatoms						
zooplankton	copepods						
	cyclops						
zoobenthos	oysters						
	chironomus						
bacteria				high molecular weight (MW)			low MW

Fig. 6.4 Principal composition of organism (simplified and modified after Huc 1980; Vandenbroucke and Largeau 2007)

Table 6.3 Biomacromolecules and their principal preservation potential in the geosphere; + high, − low preservation potential (simplified after deLeeuw and Largeau 1993)

Biomacromolecules	Occurrence	Preservation potential
Proteins, peptides	All organism	−
Glycogen	Animals	−
Starch	Vascular plants; algae; bacteria	−
Cellulose	Vascular plants; some fungi	−
Pectin	Vascular plants	−/+
DNA/RNA	All organisms	−
Chitin	Arthropods, copepods, crustacea ...	+
Cutin, subcrin	Vascular plants	+/++
Tannins	Vascular plants, algae	+++
Lignin	Vascular plants	++++
Algaenans	Algae	++++
Cutans, suberans	Vascular plants	++++

The preservation potential is primarily not related to the type of polymerization (e.g. algaenan, suberin but also peptides and polysaccharides are products of polycondensation), but the regularity seems to have an influence on the stability of polymers against microbial reworking. Both the less as well as the more organic geochemically relevant biopolymers will be introduced and discussed in the following.

General Note

Once more it becomes obvious, that the chemical structure is the key parameter influencing the environmental behavior and geochemical relevant properties of organic biomolecules. For biopolymers the regularity is a major feature for organic geochemical stability.

6.2   Polysaccharides (Carbohydrates)

Outlook

Polysaccharides represent the most abundant fraction of natural compounds in nature. They are built up of monosaccharides. Their molecular properties as well as their mode of repetitive linkage forming oligomers and finally biopolymers are introduced.

Polysaccharides are also named carbohydrates, but this name represents both low and high molecular weight substances. Polysaccharides are the result of polymerization of monosaccharides also named sugars. From a structural point of view the biological most important saccharides consist of 5 or 6 carbon atoms, which are all substituted with one hydroxyl group, respectively, with exception of one position, at which a carbonyl group is located. This can be either an aldehyde group or a keto group at C2 position. Carbon chain length and type of carbonyl group are used for nomenclature of monosaccharide sub classes. For this purposes, the term '-*ose*', generally used as suffix for monosaccharides, can be combined with a Greek numeral (e.g. hexoses and pentoses). Furthermore, the carbonyl type can be added in the same way (ketoses and aldoses) or combined with the carbon chain length information as prefix. Then, aldopentoses and ketohexoses as well as ketopentoses and aldohexoses can be differentiated (see Fig. 6.5).

The high degree of functionalization has two important implications. Firstly, several chiral carbon atoms appear. Their number depends on carbon chain length and type of carbonyl group. For example, a ketopentose exhibits two chiral centers, whereas an aldohexose has four chiral carbon atoms. As a consequence of the multiple chiral centers, different series of diastereomers exist. These classes of stereoisomers form the group of monosaccharides. However, the group of biogenic monosaccharides is restricted. An essential feature is the orientation at the next to the last carbon atom (C4 at pentoses, C5 at hexoses). This orientation can be easily illustrated using the Fischer-projection that transfers the 3-dimensional molecule to

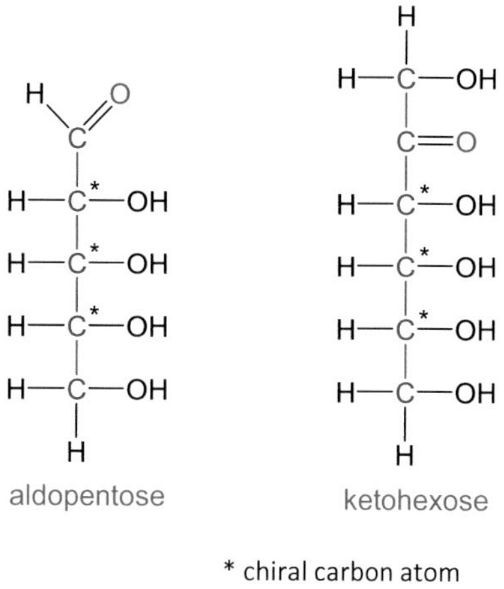

Fig. 6.5 Structural
properties of
monosaccharides

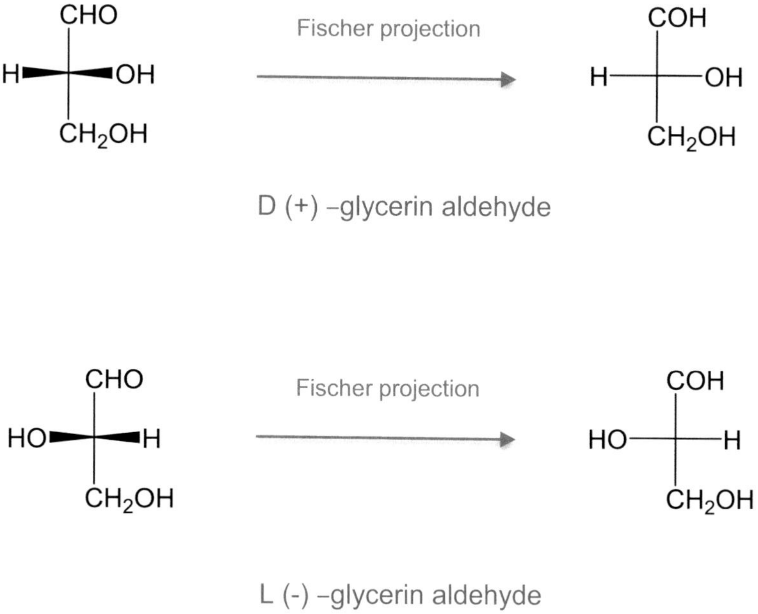

aldopentose   ketohexose

* chiral carbon atom

Fischer projection

D (+) –glycerin aldehyde

Fischer projection

L (-) –glycerin aldehyde

Fig. 6.6 The Fischer projection illustrated for glycerin aldehyde. D and L notation points to the orientation of the hydroxyl group at the middle position. The mark in the brackets (+ or −) points to the rotational direction that these optical isomers induces on plane light and is independent from D or L orientation

a 2-dimensional figure. In this projection (illustrated for the smallest chiral mono-saccharide, the C3 glycerinaldehyde in Fig. 6.6), the stereochemical properties at the carbon atoms are considered by the orientation of the hydroxyl groups (left or right).

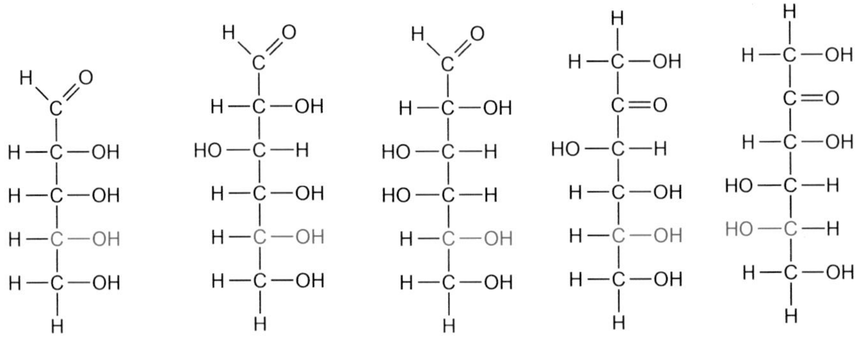

Fig. 6.7 Some examples of monosaccharides. Note the D and L system and the corresponding orientation of the hydroxyl group attached to the next to last position

Interestingly, all biogenic monosaccharides exhibit a right hand orientation of the hydroxyl group at the next to last position. For some classes of natural products a special nomenclature is used to define these individual stereoisomers using the letters D and L. The described right hand orientation is named D, whereas a left hand orientation at this specific site is named L (some examples are given in Fig. 6.7).

Hence, as mentioned all biogenic monosaccharides have a right hand orientation and, consequently, are D isomers. In summary, the group of biological relevant monosaccharides consists of all D stereoisomers of aldoses with chain lengths of 4–6 atoms (see Fig. 6.8) as well as of the corresponding ketoses (see Fig. 6.9). The most prominent representative of monosaccharides is certainly glucose but also fructose, ribose, xylose, galactose or mannose are commonly known.

A second important aspect of the multifunctionality of monosaccharides is related to the corresponding ability for multiple reactions. Some simple reactions are resulting in biomolecules such as the oxidation of glucose to gluconic acid and following to saccharic acid, or also the reduction of glucose to sorbit (see Fig. 6.10). However, these reactions do not exhibit any relevance neither for the formation of polysaccharides nor from a geochemical point of view.

Contrarily, another type of reaction has more importance for the formation of polysaccharides. A well-known reaction of carbonyls with hydroxyl groups is the two-step formation of acetals via hemi acetals as illustrated in Fig. 6.11. A first reaction step forms a half acetal by addition of a hydroxyl group at the carbonyl group, and the acetal is formed by a following condensation of a second hydroxyl group at the same carbon atom. The same reaction can be carried out by ketones forming hemi ketals and ketals. Noteworthy, all these reactions are reversible, in particular the formation of the hemi acteals and hemi ketals.

Monosaccharides uses their carbonyl group and sterically appropriate hydroxyl group for a first intramolecular reaction forming a ring system as illustrated in Fig. 6.12. The resulting ring systems have a ring size of five or six members

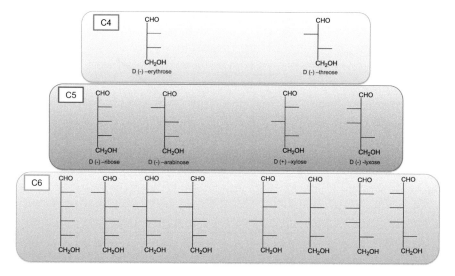

Fig. 6.8 The family of biogenic aldoses

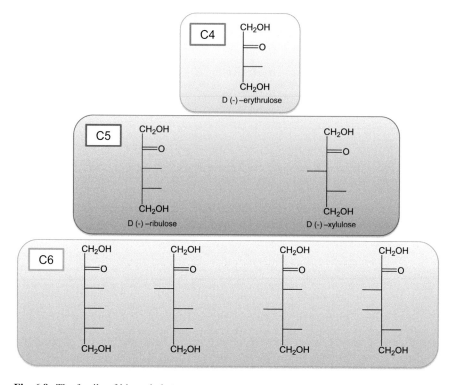

Fig. 6.9 The family of biogenic ketoses

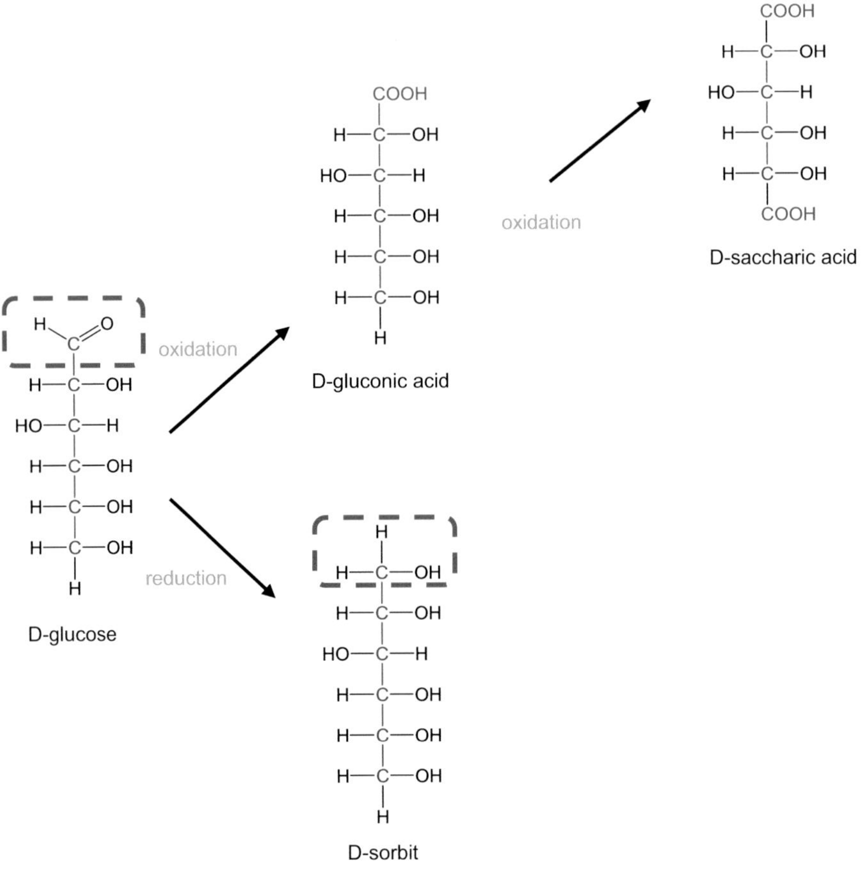

Fig. 6.10 Selected reactions of glucose

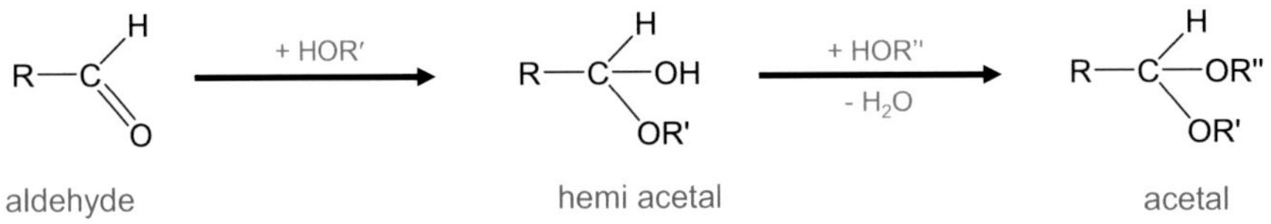

Fig. 6.11 The formation of acetals

including one oxygen atom as a result of the hemi acetal formation. Since these ring systems are chemical equivalent to the molecules tetrahydropyrane or tetrahydrofurane, these cyclic forms of monosaccharides are also called pyranoses or furanoses. The cyclisation has one important stereochemical aspect. Since the carbonyl carbon atom becomes substituted by a fourth substituent of different chemical quality, this carbon atom becomes chiral as a result of the cyclisation reaction (this carbon atom is also named anomeric center). Consequently, two

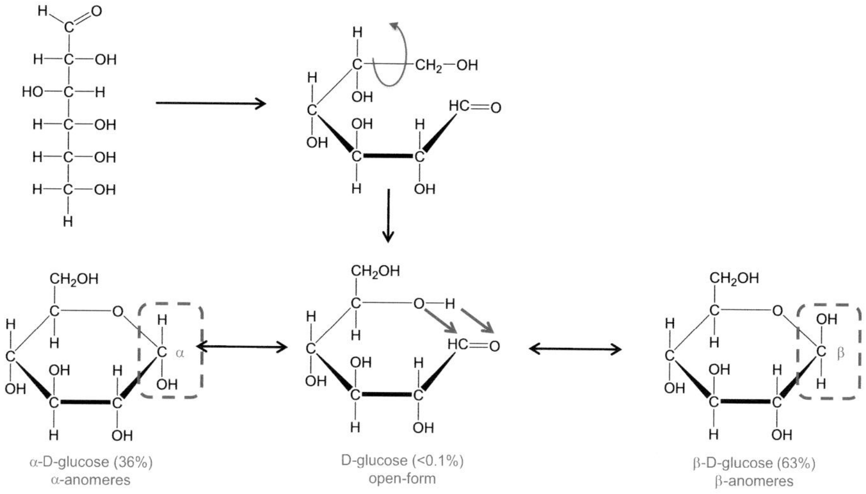

Fig. 6.12 The formation of α- and β-glucose by an intramolecular cyclisation reaction

different stereoisomers, the so-called anomers, can be formed, the α- or β-isomers. The formation of the α- and β-anomers of glucose is illustrated in more detail in Fig. 6.12. The corresponding formation of α- and β-fructose is given in Fig. 6.13. Both monosaccharides are forming pyranoses.

In order to visualize better the cyclic structure but to avoid confusing three-dimensional complexity, a special type of notation is used for cyclic monosaccharides, the Haworth depiction as illustrated in Fig. 6.14.

Since the formation of hemi acetals and ketals are reversible the open form and the ring form are in a dynamical exchange, in particular in aqueous solution. In aqueous solution the ring forms are preferred and the open form exists to a very low extent of approx. 0.1 % (see Fig. 6.12). Nevertheless, there is a very dynamic exchange between the both cyclic anomers (in average existing with an α- to β-isomer ratio of roughly 1:2) via the open form.

At this point it is not far from formation of polymers based on monosaccharides. The cyclic monosaccharides have the possibility to complete the acetal/ketal formation, but only as an intermolecular reaction using the hydroxyl group of a neighboring molecule. A corresponding linkage of two monosaccharides leads to disaccharides. Well-known disaccharides such as sucrose, maltose or lactose are exemplified in Fig. 6.15.

This type of disaccharide formation allows further condensation steps since the second hemi acetal or hemi ketal position is still chemically available for further, chain prolonging reaction steps. In this manner trisaccharides (an example is given in Fig. 6.16) and further oligosaccharides can be synthesized. And finally the multiple intermolecular reaction of saccharide hemi ketals/acetals leads to the formation of polysaccharides. This formation is very regular, since biological

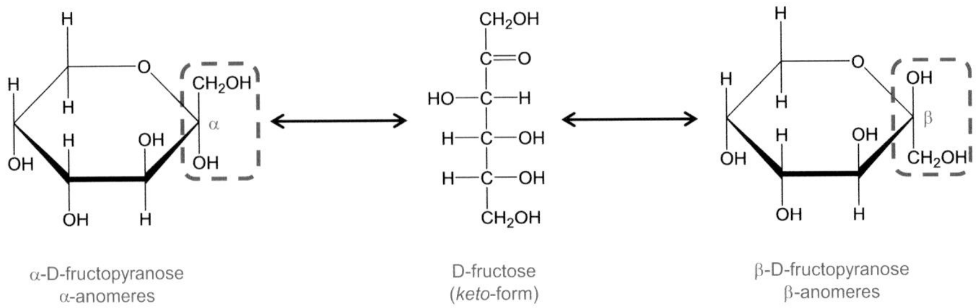

Fig. 6.13 The formation of α- and β-fructose by an intramolecular cyclisation reaction

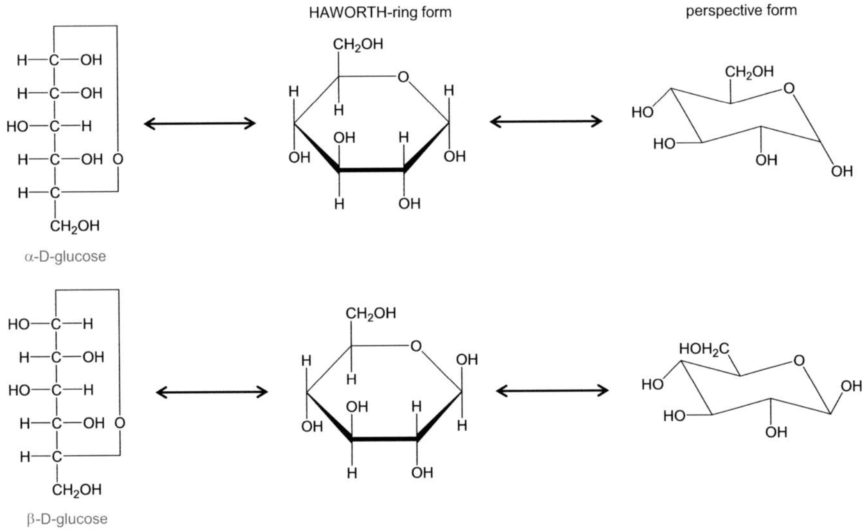

Fig. 6.14 Possibilities to plot cyclic monosaccharides

polysaccharides are based normally on only one monomer subjected to polycondensation.

Glucose acts as basic monomer for the most important polysaccharides. The continuous linkage of β-D-glucose using the 1 and 4 positions forms the most abundant biomolecule in nature, the cellulose (see Fig. 6.17). Just changing the anomers as monomeric unit forms a second important natural product, the starch. In fact starch is a mixture of two α-D glucose based polymers, the amylose and the amylopectin. Amylose is a straight polymer chain with a secondary structure similar to cellulose (also 1–4 linkage). However, amylopectin has a same skeleton, but additionally some branching with 1–6 linkages. This type of cross-linking allows swelling as a specific property of starch.

Some further polysaccharides are build up from chemically modified monosaccharides (see Fig. 6.17). Chitin is a polysaccharide formed from β-D-glucose

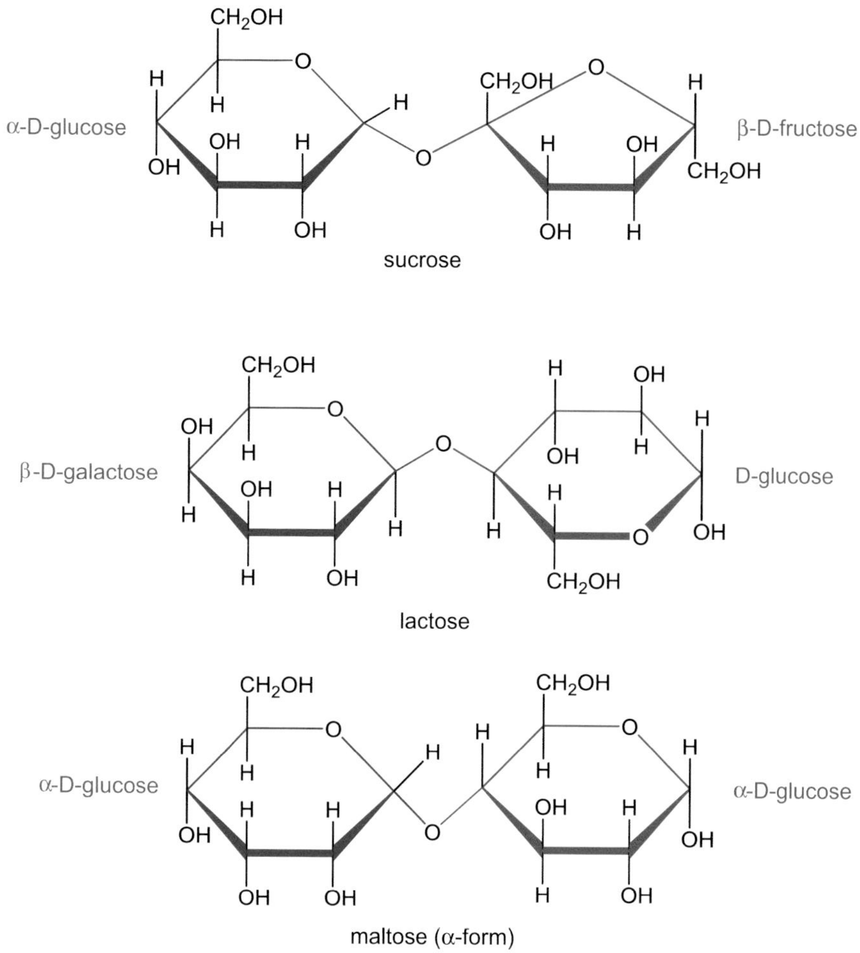

Fig. 6.15 Examples of well-known disaccharides

Fig. 6.16 Chemical structures of an exemplary trisaccharide, the antibiotic streptomycin

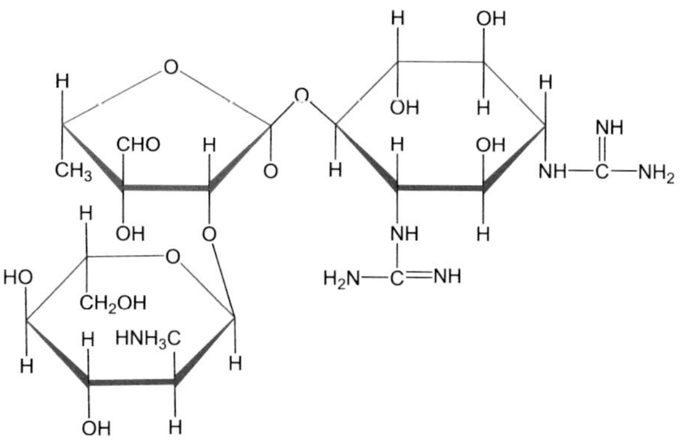

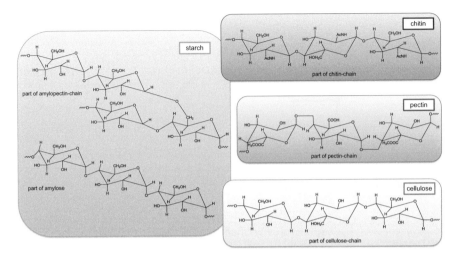

Fig. 6.17 Chemical structures of some biological important polysaccharides

Table 6.4 Composition of some polysaccharides

Name	Monomere	Bridging bond position	Occurrence
Cellulose	β-D-glucose	1–4	Plants
Amylose	α-D-glucose	1–4	Plants
Amylopektin	α-D-glucose	1–4 and 1–6	Plants
Glycogen	α-D-glucose	1–4 and 1–6	Animals
Dextran	α-D-glucose	1–6, 1–3, 1–4, 1–2	Bacteria
Chitin	β-D-aminoglucose	1–4	Insects, crabs, fungi
Xylan	β-D-Xylose	1–4	Plants

monomers, in which one hydroxyl group is exchanged by an acetylated amino group. Pectin exhibits methylated carboxylic groups in the polysaccharide chain.

Generally, the individual polysaccharides are characterized by both their monomers and the position of their bridging linkages. Examples are summarized in Table 6.4.

As mentioned in the former sub chapter polysaccharides are less stable in the geosphere and, therefore, exhibit a very low preservation potential. Consequently, the application of polysaccharide analyses in Organic Geochemistry is scarce and limited in the characterization of immature organic matter. However, polysaccharides exhibit the potential to differentiate organic matter sources to some extent. For this purpose the composition of the polysaccharide fraction of a given sample can

Fig. 6.18 Saccharide based
approach to discriminate
different types of plant
material (modified after
Killops and Killops 2005;
Cowie and Hedges 1984)

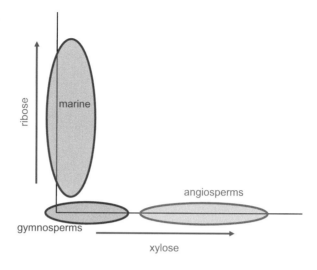

be used to distinguish the type of plant organic matter. Source specific properties
are allocated to ribose, xylose, arabinose and galactose. Correlation of individual
contributions of these monosaccharides has been used for differentiating angiosperms from gymnosperms and woody from non-woody material as exemplified in
Fig. 6.18.

Case Example: Chitin in the fossil record

Chitin is one of the more resistant polysaccharides. Hence, it can be identified
also in fossil matter to some extend. Due to its polymer structure, it has been
analyzed dominantly by invasive methods, like pyrolysis GC/MS or chemical
degradation. Applying such degrading methods on organic matter leads to a
break down of the polymeric structure and the low molecular weight products
can be easily analyzed. If these products are specific for the macromolecular
precursor a more or less unambiguous identification is possible. Using this
approach, chitin has been detected in arthropod fossils from the Miocene,
Pliocene, Pleistocene and even Miozene, representing a record of up to
25 Mio years (Flannery et al. 2001).

(continued)

Figure: Characteristic pyrolysis and hydrolysis products obtained from analyses of fossil insects (according to Flannery et al. 2001)

6.3   Proteins and Peptides

Outlook

In a similar way as already described for polysaccharides, also peptides and proteins are regularly built up by monomers, in this case the amino acids. The spectrum of proteinogenic amino acids, their polycondensation and their resulting structural features are discussed.

Peptides and proteins are less abundant in nature but have also a high biological importance. They do not differ generally by their chemical composition but by their molecular weights. Peptides can be classified as small proteins.

The systematic description of these macromolecules can follow the same way the polysaccharides have been introduced. Firstly, the monomeric units building up later the macromolecular structure need to be characterized. Basic components of proteins are amino acids, in more detail α-amino acids. This functional group induces a high chance for chirality at the α-carbon atom. If substituents different to H or COOH appear at this position, enantiomers are formed, which are

$$\text{COOH}$$
$$\text{H}_2\text{N}\!-\!\overset{|}{\underset{|}{\text{C}}}\!-\!\text{H}$$
$$\text{R}$$

α-amino acid

CH$_2$OOH

H$_2$N►C◄H

R

fischer projection

CH$_2$OOH

H►C◄NH$_2$

R

L-form

D-form

Fig. 6.19 Principal chemical structure and stereoisomer denomination of amino acids

denominated by L or D according to the Fischer projection (see Fig. 6.19). In nature L amino acids occur more or less exclusively.

Natural occurring amino acids with relevance for proteins are a limited group of 20 so-called proteinogenic amino acids that are summarized in Fig. 6.20. They can be categorized either by their chemical characteristics (dicarboxylic acids, aromatic moities, cyclic structures, sulphur containing substances etc.) or by the possibility for humans to biosynthesize the individual amino acids. Essential amino acids are those derivatives that humans are not able to synthesize *ab inito* and, therefore, needed to be supplied by food.

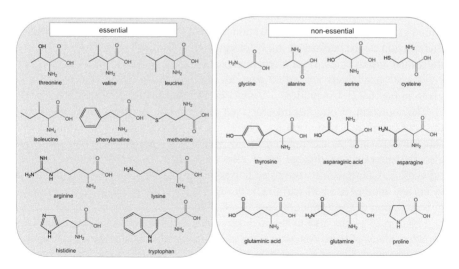

Fig. 6.20 All proteinogenic amino acids separated into essential and non-essential substances

Table 6.5 Exemplary proteinogenic amino acids with their shortcuts. For some derivatives a linkage of names with occurrence or properties exist

Name	Shortcut	Origin of name	Essential
Mono-amino mono-carboxyl acids			
Alanine	Ala		–
Leucine	Leu		✓
Phenylalanine	Phe		✓
Serine	Ser	Lat. sericum = silk	–
Valine	Vai	Lat. validus = healthy	✓
Di-amino mono-carboxyl acids			
Arginine	Arg	Lat. argentum = argent	✓
Mono-amino di-carboxyl acids			
Asparigine acid	Asp		–
Glutamine acid	Gin	Lat. glutinum = glue	–
Heterocyclic-amino acids			
Histidine	His	Greek histos = histoid	✓
Proline	Pro		–
Tryptophane	Try	Trypsine + Greek Phainein = emerge	✓

For a simplified annotation in particular of proteins the proteinogenic amino acids are characterized by three letter shortcut as given for some examples in Table 6.5.

Since amino acids are multiple functionalized they can react in different ways. Some relevant reactions are summarized in Fig. 6.21. Amino groups can react as alkali and carboxylic acids exhibit acidic properties, hence both functional groups can react intramolecular in a classical acid–base reaction. In aqueous solution this reaction is influenced by the pH value. The pH value, at which both groups are charged resulting in a non-charged molecule, is called the isoelectronic point and is specific for each individual amino acid.

For building up biopolymers the second reaction type is relevant. Carboxylic and amino groups can react in a condensation reaction forming amides. For amino acids this reaction can be performed solely intermolecular leading to a multiple linkage of individual amino acids, from oligocondensation up to polycondensation. The first condensation products are called peptides comprising dipeptides (an example is given in Fig. 6.22), tripeptides, ..., oligopeptides and finally polypeptides.

The depiction of oligopeptides or even polypeptides becomes confusing using the traditional structure formula. Therefore, the already mentioned three-letter shortcuts are used as exemplified in Fig. 6.23. Note that not only the order but also the direction of linkage is of relevance since the amide bonds have an orientation. Consequently, oligo- or polypeptides exhibit a strain end with amino group and another end with carboxylic group. These different end groups are named N-termination or C-termination (see Fig. 6.21).

Figure 6.23 points also to the high physiological relevance of oligo- and poly-peptides. Many of these compounds act as diverse regulators in organism. An interesting structural feature is related to the amino acid cysteine exhibiting a sulphur containing functional group. This thio group is used in peptides to form intramolecular (see oxytocin in Fig. 6.23) but also intermolecular sulfide bridges. The latter one enables a linkage e.g. of two different peptide strains as illustrated in Fig. 6.24.

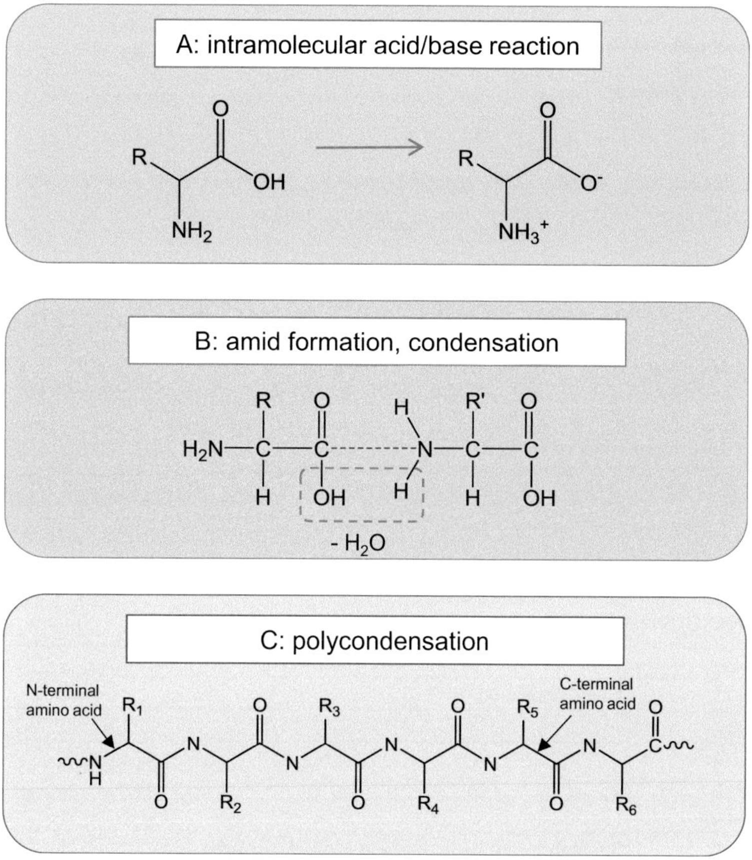

A: intramolecular acid/base reaction

B: amid formation, condensation

$-H_2O$

C: polycondensation

N-terminal amino acid R_1

C-terminal amino acid

R_3

R_5

R_2

R_4

R_6

Fig. 6.21 Reaction types of amino acids

Fig. 6.22 Aspartam as an example for a dipeptide (built up by phenylalanine methylester and asparaginic acid)

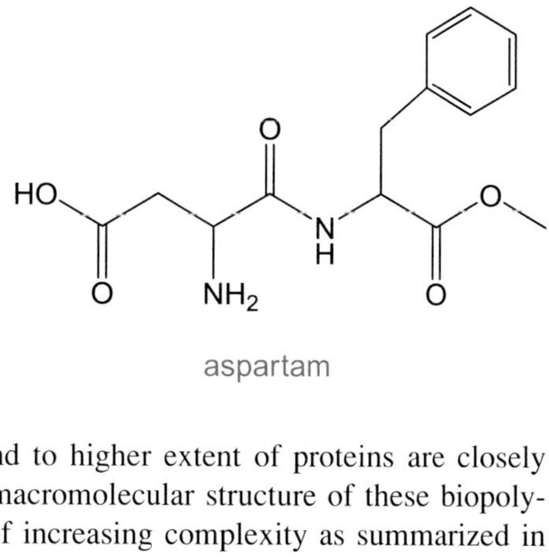

aspartam

Physiological effects of peptides and to higher extent of proteins are closely related to their overall structure. The macromolecular structure of these biopolymers is described on different levels of increasing complexity as summarized in Table 6.6.

The primary structure is just the order or sequence of amino acids in the polymers. How the individual chains are arranged is described by the secondary

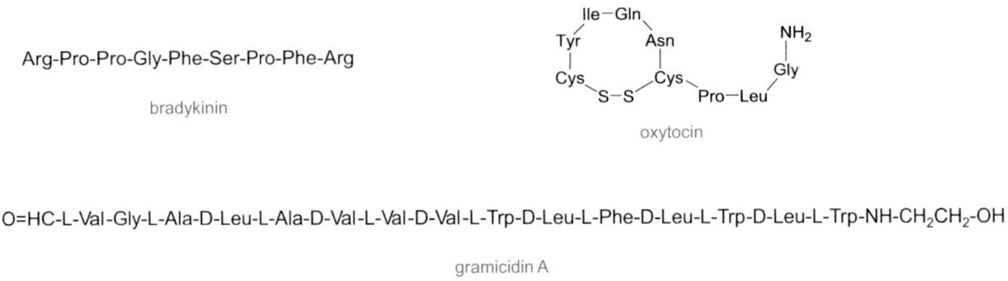

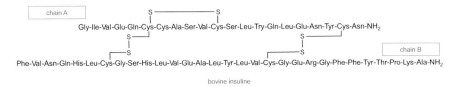

bradykinin

Arg-Pro-Pro-Gly-Phe-Ser-Pro-Phe-Arg

oxytocin

O=HC-L-Val-Gly-L-Ala-D-Leu-L-Ala-D-Val-L-Val-D-Val-L-Trp-D-Leu-L-Phe-D-Leu-L-Trp-D-Leu-L-Trp-NH-CH₂CH₂-OH

gramicidin A

Fig. 6.23 Examples of physiological relevant oligopeptides (bradykinin = inflammatory mediator; oxytocin = mammalian hormone; gramicidin A = antibiotic from soil microbes)

chain A

Gly-Ile-Val-Glu-Gln-Cys-Cys-Ala-Ser-Val-Cys-Ser-Leu-Try-Gln-Leu-Glu-Asn-Tyr-Cys-Asn-NH₂

Phe-Val-Asn-Gln-His-Leu-Cys-Gly-Ser-His-Leu-Val-Glu-Ala-Leu-Tyr-Leu-Val-Cys-Gly-Glu-Arg-Gly-Phe-Phe-Tyr-Thr-Pro-Lys-Ala-NH₂

chain B

bovine insuline

Fig. 6.24 Insulin of cows as example for intermolecular sulfide bridges in peptides

Table 6.6 Overview on different structural levels of proteins

Proteins	
Structure	Characterisation
Primary	
Amino acid sequence	
Secondary	
α-Helix	3,6 amino acid parts per twist = 5.4 Å per twist
β-Sheet	Neighbouring chains run in opposing direction & held together at hydrogen bridge bonds (7 Å)
Tertiary	
Spheroproteins	Myoglobin – oxygen carrier in muscles
Fibrous protein	Creatine – component of skin, hair, nails, feathers Collagen – conjunctive tissue, tendons, bones, cartilage
Quaternary	
Aggregate formation	Haemoglobin – oxygen carrier in the bloodstream

structure. Principally, the chains are forming α-helices (by two individual macromolecules) or are arranged by layers with parallel orientation of the chains on opposite direction. All these so-called β-sheets are stacked with a minor attraction between individual layers as compared to the interactions between individual chains within the sheets (see Fig. 6.25).

The last two structural features describe the overarching construction of helices and sheets. The tertiary structure differentiates spheroproteins and scleroproteins. Scleroproteins (also called fibrous proteins) are essential constituents of various organs, e.g. collagen as a main component of bones or creatine forming nails or hair

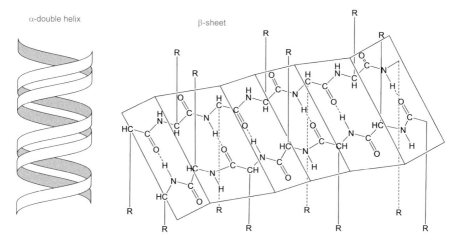

Fig. 6.25 Secondary structures of proteins

(see Table 6.7). On the contrary, spheroproteines exhibit regulatory functions like enzymes or carriers (see Table 6.8). Lastly, the quaternary structure describes the overall three dimensional arrangement of the proteins and linkages to further non-protein moieties (like saccharides, metal ions etc.) or aggregates.

Excursus: *Differences between a perm and a blow-dried hairdo as explained by protein chemistry*

Protein chemistry is the clue for understanding some aspects of *hair do*. In detail ceratine as important constituent of hair fibers is responsible for forming some types of hair dos. Wetting hairs leads to flexibility due to cleavage of hydrogen bonds of the proteins. The hair can be formed and fixed again by drying. However, the stability of such blow-dried hairdo is limited. As soon as the hair gets wet again, the blow-dried hairdo is lost.

A higher stability can be obtained by cleaving and linking another type of intramolecular protein linkages, the disulfide bridges. As a first step this sulfide bonds between cysteine moieties are cleaved by reduction using thioglycolic acid salts. After forming the hair, a second reaction step, an oxidation, relinks the cysteine moieties via new disulfide bridges using hydrogen peroxide. Since this stabilization is a based on covalent bonds, the durability of a perm is much higher.

$$\text{Cys-S-S-Cys} + 2\ \text{HS-CH}_2\text{-COO}^-\ \text{NH}_4^+ \longrightarrow \text{Cys-SH} + \text{HS-Cys} + \text{NH}_4^+\ {}^-\text{OOC-CH}_2\text{-S-S-CH}_2\text{-COO}^-\ \text{NH}_4^+$$

$$\text{Cys-SH} + \text{HS-Cys} + \text{H}_2\text{O}_2 \longrightarrow \text{Cys-S-S-Cys} + 2\ \text{H}_2\text{O}$$

Table 6.7 Examples of scleroproteins and their function in organism

Example	Function
Creatine	Component of skin, hair, nails, feathers
Collagen	Conjunctive tissue, tendons, bones, cartilage
Fibroin	Gossamer, silk cocoon
Sclerotin	Insect shells
Myosin	Contractile component of muscles
Actin	Contractile component of muscles

Table 6.8 Examples of spheroproteins and their function in organism

Example	Type	Function
Carboxypeptidase	Enzyme	Catalysis of polypeptide chain hydrolysis
Trypsin	Enzyme	Catalysis of polypeptide chain hydrolysis
Haemoglobin	Carrier	Oxygen carrier in the bloodstream
Myoglobin	Carrier	Oxygen carrier in muscles
Cytochrome	Carrier	Electron transfer
Ovalbumin	Storage	Food storage in proteins
Casein	Storage	Milk protein
Antibodies	Protection	Formation of insoluble Complexes with foreign bodies
Insulin	Hormone	Regulation of glucose metabolism

Although peptides and proteins play an essential role in all organism, their value for Organic Geochemistry is very limited due to their very low persistence after release to the geosphere. The overall balance of the amino acid budget in aquatic systems points to a very small fraction of amino acids that survives the consumption and degradation processes in the water column and enters the sediments, where further rapid degradation occurs. In summary, although peptides and proteins exhibit very high biological specifity, their decomposition is too fast to observe them over geological time scales. There is only one very special application of amino acid analyses in Organic Geochemistry that is related to the stereochemical properties at the α-carbon atom. As already mentioned, biogenic amino acids exhibit the L-configuration. However, an epimerization can be observed as the result of intramolecular rearrangement with a planar carbanion as state of transition (as illustrated in Fig. 6.26).

Since these reaction needs a prolonged time range the relative proportion of biogenic L-amino acids as compared to the D-isomer can be used for dating issues. Ranges of a linear relationship between the D/L ratio and time have been observed as exemplified in Fig. 6.27. Noteworthy, free (extractable) and bound (non-extractable) amino acid epimerization have different time scales, but can be used complementary for dating sediment cores with ages of up to approx. one hundred thousand years.

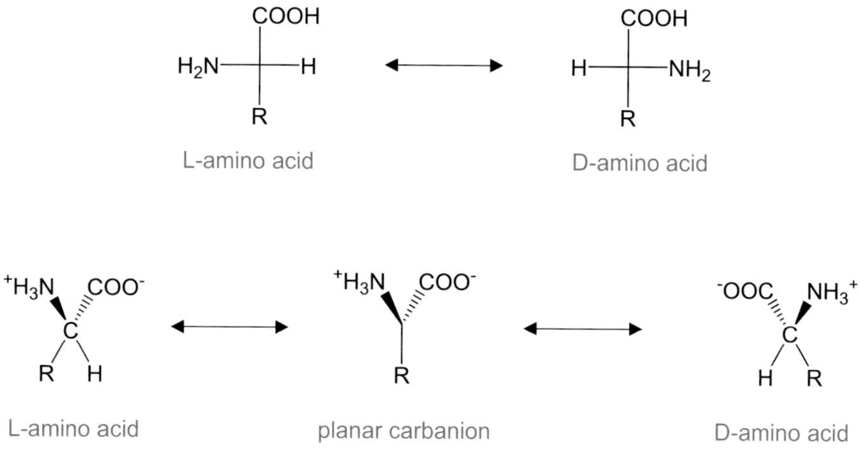

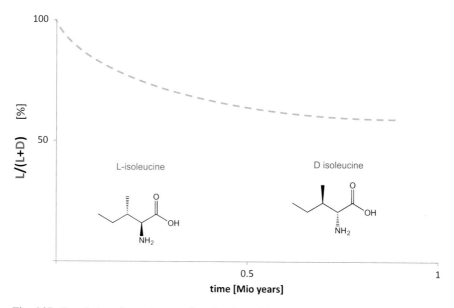

Fig. 6.26 Reaction scheme of intramolecular epimerization of amino acids

Fig. 6.27 Correlation of enantiomer ratios of amino acids and age of sediments (simplified and modified after Kaufman and Miller 1992)

6.4   Suberin, Cutin, Sporopollenin and Algaenan

Outlook

Besides regular biopolymers there are some geochemical more relevant biopolymers with irregular structure. On the one hand these macromolecules are
specifially relevant for distinctive funtions in plants, on the other hand they
exhibit a high preservations potential. Well known examples comprising
cutin, suberin, sporopollenin and algaenan are introduced.

In contrast to the already discussed more regular biopolymers, some biological
macromolecules with significant higher preservation potential in the geosphere
have a more irregular constitution. Suberin, cutin, sporopollenin and algaenan
formations are based on a spectrum of polyfunctionalized monomers which can
react by various reaction types. Crosslinking allows building up a more resistant
molecular network and together with the randomized forming of linkages the major
reason for the high recalcitrance.

Cutin and suberin are basic constituents in plant surfaces (see Fig. 6.28). Cutin is
part of plant cuticle waxes and contributes to the exchange regulation and transport
of water and gases from and to the leaf interior. The same function exhibits suberin,
but not as ingredient of leaf cuticles but of roots.

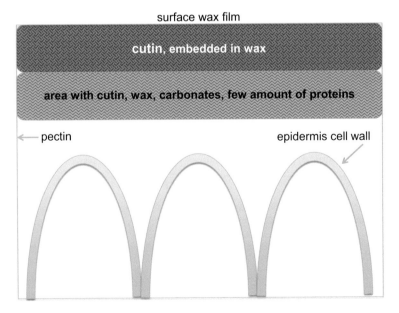

Fig. 6.28 Cutin in leaf cuticles

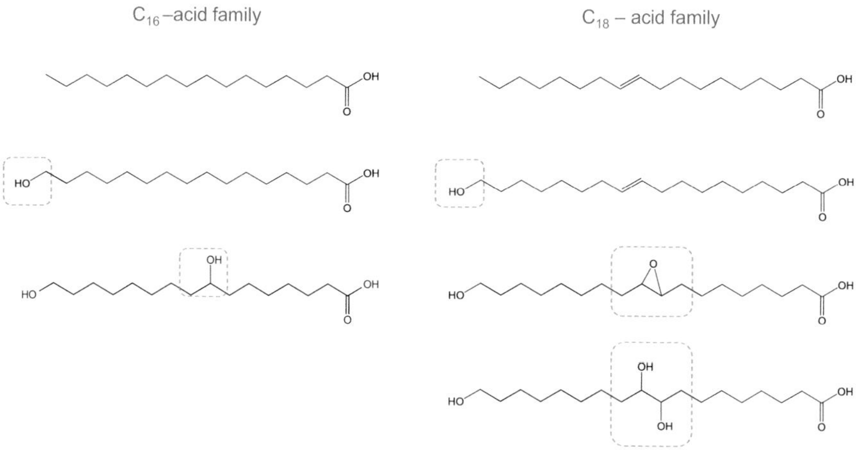

C$_{16}$–acid family   C$_{18}$ – acid family

Fig. 6.29 Basic structural elements of suberin and cutin (according to Kolattukudy 1980)

The molecular composition of both, cutin and suberin, is highly similar. Functionalized fatty acids represent the major type of monomers (see. Fig. 6.29). The dominant type of linkage is an ester condensation, hence cutin and suberin can be characterized to some extent as polyesters. Slight differences exist for the basic components with respect to chain length and type as well as degree of functionalization. Basic units are C16 and C18 carboxylic mono and di acids with hydroxyl or epoxy groups. Also alcohols, either long chain aliphatic or aromatic ones, contribute to the macromolecular structure.

Structural differences between cutin and suberin can be emphasized by the monomer composition as summarized in Table 6.9. As an example a somewhat higher degree of aromatization is observed for suberin, whereas cutin exhibit more in chain substituted fatty acid moieties (Fig. 6.30).

Case Example: Cutin and suberin in soil science

The significance of some components to distinguish between cutin and suberin is used in soil sciences to follow the fate of plant material in soil. Characteristic moieties obtained after alkaline hydrolyses have been used by Otto and Simpson (2006) to estimate the input of root and leave material to soils. As exemplified below.

(continued)

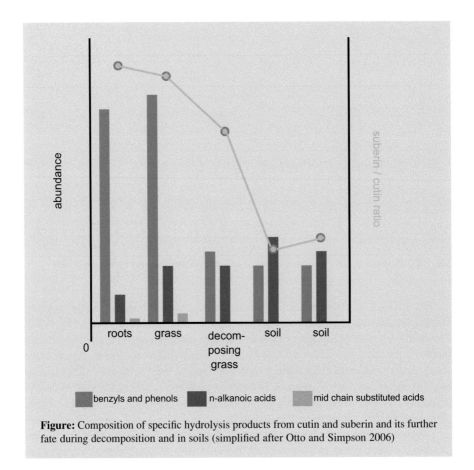

Figure: Composition of specific hydrolysis products from cutin and suberin and its further fate during decomposition and in soils (simplified after Otto and Simpson 2006)

Structurally high similarity is also obvious for two further biomacromolecules – algaenan and sporopollenin. They are built up from functionalized fatty acids as well and exhibit a cross-linked molecular network dominantly formed by ester and ether linkages. Algaenan represents the resistant material in cell walls of algae. It is composed of long chain and unbranched aliphatic moieties and ether cross linkages. A structural proposal is given in Fig. 6.31.

Sporopollenin is a fourth irregular biopolymer and an essential part of the outer cell wall of spores. It is consisting basically of functionalized fatty acids, amino acids, some phenolic units and further components building up a relatively resistant chemical network. Due to its recalcitrance against microbial reworking sporopollenin has been detected in many sedimentary systems.

Table 6.9 Basic monomeric units of cutin and suberin obtained after alkaline hydrolyses (adapted and simplified from Kolattukudy 1980 and references cited therein)

Monomer class	Range/type	Cutin	Suberin
Phenols	Coumaryl, vanillyl, syringyl	Minor	Major
n-alkanols	C_{16}–C_{32}	Minor	Major
n-alkanoic acids	C_{12}–C_{36}	Minor	Major
Long chain ω-hydroxyalkanoic acids	C_{20}–C_{32}	Minor	Major
Mid chain hydroxy acids	C_{14}, C_{15}, C_{17}	Major	Minor
Mono and dihydroxy acids and diacids	C_{16}	Major	Minor
Short chain ω-hydroxy acids	C_{16}, C_{18}	Major	Major
Di-and trihydroxy acids	C_{18}	Major	Major
Epoxy hydroxy acids	C_{18}	Major	Major
Very long chain acids	C_{20}–C_{26}	Minor and rare	Major and substantial
Very long chain alcohols	C_{20}–C_{26}	Minor and rare	Major and substantial
Dicarboxylic acids	–	Minor	Major

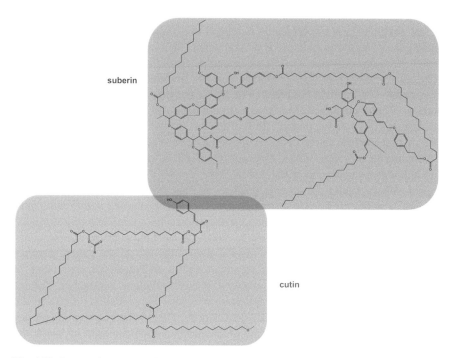

Fig. 6.30 Proposed structural elements of suberin and cutin

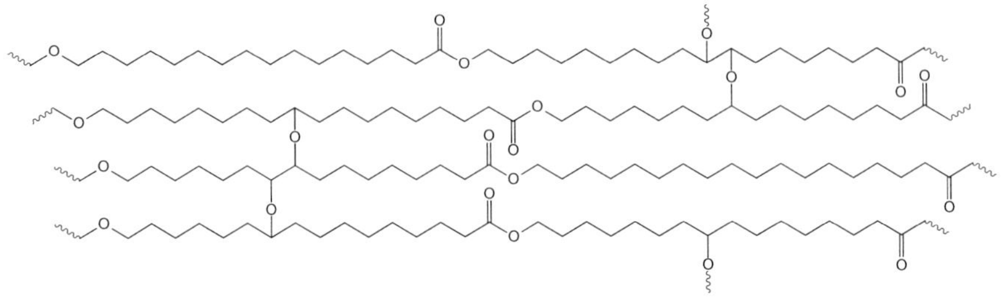

Fig. 6.31 Proposed chemical structure units in algaenan (according to Gelin et al. 1997; Blokker et al. 1998)

6.5   Lignin

Outlook

Lignin as important biopolymer in higher land plants is closely linked with wood. Its chemical properties and the geochemical relevance, as well as its analytical characterization is discussed.

A last irregular biomacromolecule is related to a unique property of higher land plants, the potential to build up rigid structures. This feature is based on wood or woody material of which a major and essential constituent is lignin. Lignin is the third most abundant organic substance on earth (beside cellulose and chitin). The macromolecule lignin is produced in plants by enzymatic polymerization of up to three monomer components with very similar chemical constitution. The simplest monomer is *p*-coumaryl alcohol, from which the further two components (coniferyl and sinapyl alcohol) can be deduced by addition of one or two methoxy substituents in *m*-position (see Fig. 6.32). Noteworthy, lignin is the most abundant natural product with aromatic moieties.

For polymerization the double bond as well as the hydroxyl groups are used forming a randomized cross-linked network. Lignin is composed in different plants by different contributions of the individual monomers. For example in conifers lignin is produced more or less exclusively by coniferyl alcohol, angiosperms use dominantly sinapyl and coniferyl alcohol and grasses often use all three monomers for lignin synthesis. These differences are also used for analytical characterization of lignin or lignin moieties e.g. in soil material or terrestrial humic substances. Clue for this analytical approach was the application of a selective oxidation procedure, the so-called copper oxide oxidation (CuO oxidation). This method cleaves the macromolecular compounds producing low molecular oxidation products in which the substitution pattern at the benzene ring persists. The relative amount of these indicative lignin oxidation products can be used to distinguish between angio- and gymnosperms as well as between woody and non woody plant material. In detail, the *p*-hydroxy-*m*-methoxyphenyl (guaiacyl), *p*-hydroxy-*m,m*-dimethoxyphenyl (syringyl) containing oxidation products as well as substances with a cinnamyl group are used for a detailed differentiation (see Fig. 6.33).

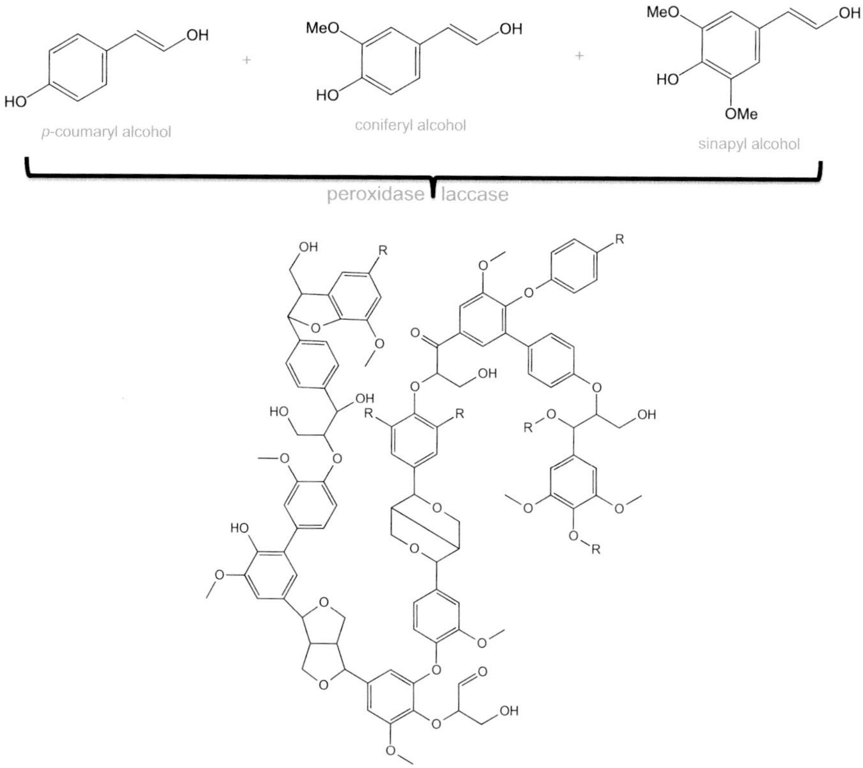

Fig. 6.32   Lignin biosynthesis and exemplary molecular structure

The two dimensional correlation of the most indicative parameters (guaiacyl and syringyl moieties) illustrates the enrichment of partial structural units and the corresponding potential to differentiate plant material based on lignin oxidation products (see Fig. 6.34).

group (chemical structure)	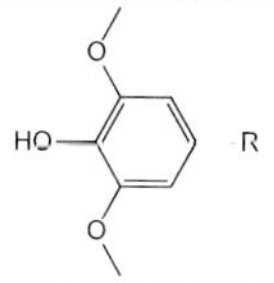 syringyl	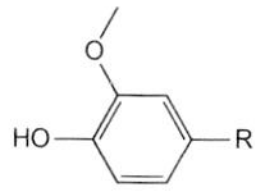 guaiacyl	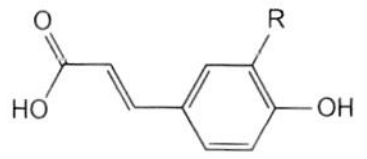 cinnamyl
non vascular plant	0	0	0
non woody angiosperms	1 -3	0.6 – 3.0	0.4 – 3.1
woody angiosperms	7 -18	2.7 – 8.0	0
non-woody gymnosperms	0	1.9 – 2.1	0.8 – 1.2
Woody gymnosperms	0	4 -13	0

Fig. 6.33   CuO oxidation products of lignin and their relative abundance (wt %) in different plant material (compiled and modified after Killops and Killops 2005; Hedges and Mann 1979)

Fig. 6.34 Correlation of indicative lignin oxidation products for differentiation of plant material (simplified after Killops and Killops 2005; Hedges and Mann 1979)

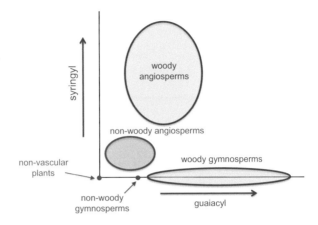

The application of this approach is limited, since diagenetic and catagenetic processes induce chemical alterations that modifies the indicative moieties in lignin derived material. In particular defunctionalization changes the specific substitution pattern at the phenyl rings by demethylation (changing methoxy to hydroxy groups) or demethoxylation/dihydroxylation (removing the substituents at the phenyl rings). Further on, cyclisation and ongoing aromatization also alters substantially the original molecular lignin finger-print (see Fig. 6.35). Hence, the CuO oxidation approach is limited to a more immature material.

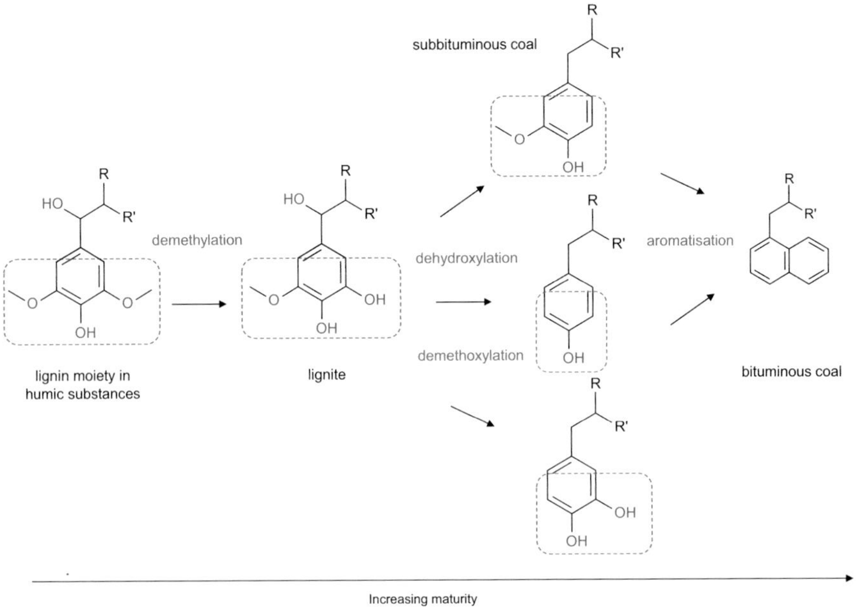

Fig. 6.35 Chemical alteration of lignin derived moieties in coaly organic matter at different maturity

6.6   Irregular Biopolymers and Their Impact on Fossil Organic Matter

All mentioned more irregular biopolymers play an important role in building up fossil material. They contribute to kerogen, coals and related matter in varying amounts. Based on their distinct resistance against microbial reworking or degradation in sedimentary systems, they remain partially unaffected in fossil matter. Detailed knowledge about chemical transformations as the result of diagenesis or early catagenesis is very limited due to their nondistinctive chemical structure. Diagenetic overprint is encompassed by the concept of humic substances as the diagenetic intermediates between biopolymers and mature macromolecular organic matter. The chemical composition of humic substances is known to comprise a wide variety depending on the origin (e.g. aquatic/marine vs. terrestrial) and the conditions during early diagenesis. For example, terrestrial humic substances exhibit higher degree of aromatic units as the result of lignin contributions. This has been shown *inter alia* by the above introduced copper oxide based oxidation approach (Ertel et al. 1984). Further on, the mechanism of humic substances formation at the early stage is still under discussion and will not followed in detail here. Principally, either the chemical conversion of existing macromolecular bio-molecules (polysaccharides, peptides, cutin, suberin, lignin …) and further crosslinking or the decomposition of macromolecules (forming low to medium molecular weight products) and a following repolymerization or aggregation is assumed to form primary humic substances. The debate is still in progress (see e.g. Sutton and Sposito 2005), but the importance of irregular biopolymers in particular for the resistant fraction of humic substances is obvious. Finally, just some general trends in chemical modification are known in course of diagenesis and catagenesis following the principal mechanisms of defunctionalisation and aromatization.

Excursus: *Classification of humic substances*

Humic substances have been classified primarily according to their water solubility at different pH values. The fraction insoluble in water at any pH is called humins, whereas those fractions how are soluble either at any pH or solelyunder acidic conditions are classified as fulvic or humic acids, respectively. However, beside this technical differentiation, some chemical differences (e.g. frequency of acidic functional groups, molecular weight, aromaticity, oxygen content) correlate well with diagenetic changes and, therefore, the three fraction represent also different stages of degradation.

(continued)

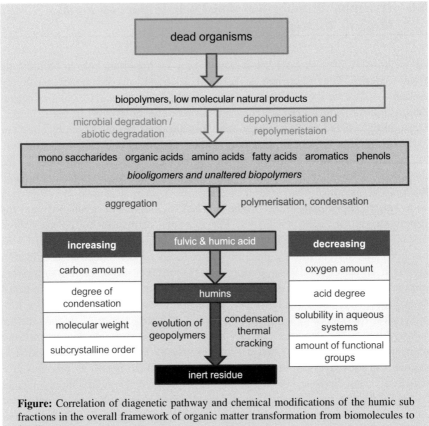

Figure: Correlation of diagenetic pathway and chemical modifications of the humic sub fractions in the overall framework of organic matter transformation from biomolecules to fossil matter

However, due to the molecular network of most irregular biopolymers as result of extended cross linking the three dimensional constitution or shape of the corresponding cell parts can survive especially in micro particles. Under favorable conditions this allows the preservation of cell components such as cell walls. Such micro particles are studied by microscopy and a rough correlation of corresponding macerals and chemofossils of biopolymers can be observed. E.g., the resistant algae material, the algaenan, correlates well with corresponding maceral alginate representing fossil residues of algae cell walls as revealed by microscopic analyses. Same correlations are evident for sporopollenin and sporinite, visible residues of spore cells, as well as for cutin and cutinite representing leaf cell residues (see Table 6.9).

Chemofossils derived from irregular biopolymers have a further important property that is related to the potential to release low molecular fragments as the result of catagenetic processes. During catagenesis high temperatures and pressure induce cracking reactions of fossil macromolecular organic matter (e.g. kerogen)

Table 6.10 Resistant irregular biopolymers and their correlation with macerals and catagenetic cracking products (adapted and modified from de Leeuw and Largeau 1993 and references cited therein)

Resistant biopolymers	Macerais	Examples of major cracking/catagenetic products
Algaenan	Alginite	*n*-alkanes, aromatics
Cutin	Cutinite	*n*-alkanes
Suberin	Suberinite	*n*-alkanes, aromatics
Lignin	Vitrinite	(poly)aromatics
Sporopollenin	Sporinite	*n*-alkanes, (poly)aromatics

generating crude oil. The quality of the fossil biopolymers determines the chemical composition of the produced petroleum and, consequently, the chemical properties of the biopolymers define as a main parameter (besides chemical changes as the result of diagenetic and catagenetic processes) the constitution of the fossil macromolecular matter (see Table 6.10). As a simple example, huge contributions of long chain unbranched aliphatic moieties in irregular biopolymers (like in algaenan) results in high aliphatic content in corresponding macromolecular chemofossils and the organic matter composed of them. Under appropriate conditions (elevated temperature and pressure – the oil window) long chain *n*-alkanes are released as the result of cracking and, consequently, the produced crude oil will consist of heightened proportions of *n*-alkanes. Hence, knowledge about the quality of biomacromolecules is a precondition for appraising the quality of fossil matter.

General Note

The diagenetic fate of more irregular biomacromolecules opens a wider view on the general composition of the most abundant form of organic matter, the kerogen. Although chemically altered some structural features survive and link biogenic origin and fossil matter properties.

References

Blokker P, Schouten S, van den Ende H, de Leeuw JW, Hatcher PG, Sinninghe Damste JS (1998) Chemical structure of algaenan from the fresh water algae *Tetraedron minimum, Scenedesmus communis* and *Pediastrum boryanum*. Org Geochem 29:1453–1468

Cowie GL, Hedges JI (1984) Carbohydrate sources in a coastal marine environment. Geochim Cosmochim Acta 48:2075–2087

De Leeuw JW, Largeau C (1993) A review of macromolecular organic compounds that comprise living organisms and their role in kerogen, coal, and petroleum formation. In: Engel MH, Macko SA (eds) Organic geochemistry – principles and applications. Plenum Press, New York/London, pp 23–72

Ertel JR, Hedges JI, Perdue EM (1984) Lignin signature of aquatic humic substances. Science 223:485–487

6 Macromolecules bibliography page 160

Flannery MB, Stott AW, Briggs DEG, Evershed RP (2001) Chitin in the fossil record: identification and quantification of D-glucosamine. Org Geochem 32:745–754

Gelin F, Boogers I, Noordeloos AAM, Sinninghe Damste JS, Riegman R, de Leeuw JW (1997) Resitant biomacromolecules in marine microalgae of the classes Eustigmatophyceae and Chlorophyceae: geochemical implications. Org Geochem 26:659–675

Hedges JI, Mann DC (1979) The characterization of plant tissues by their lignin oxidation products. Geochim Cosmochim Acta 43:1803–1807

Huc AY (1980) Origin and formation of organic matter in recent sediments and its relation to kerogen. In: Durand B (ed) Kerogen, insoluble organic matter from sedimentary rocks. Editions Technip, Paris, pp 445–474

Killops S, Killops V (2005) Introduction to organic geochemistry, 2nd edn. Blackwell Publishing, Oxford

Kolattukudy PE (1980) Cutin, suberin, and waxes. In: Stumpf PK (ed) The biochemistry of plants, vol 4. Academic Press, New York, pp 571–645

Otto A, Simpson MJ (2006) Sources and composition of hydrolysable aliphatic lipids and phenols in soils from western Canada. Org Geochem 37:385–407

Sutton R, Sposito G (2005) Molecular structure in soil humic substances: the new view. Environ Sci Tech 39:9009–9014

Vandenbroucke M, Largeau C (2007) Kerogen origin, evolution and structure. Org Geochem 38:719–833

Further Reading

Bada JL, Schroeder RA (1975) Amino acid racemization reactions and their geochemical implications. Naturwissenschaften 62:71–79

Kaufman DS, Miller GH (1992) Overview of amino acid geochronology. Comp Biochem Physiol 102B:199–204